SCIENCE AND TECHNOLOGY
IN MEDIEVAL SOCIETY

ANNALS OF THE NEW YORK ACADEMY OF SCIENCES

Volume 441

SCIENCE AND TECHNOLOGY
IN MEDIEVAL SOCIETY

Edited by Pamela O. Long

The New York Academy of Sciences
New York, New York
1985

Cover art: *The saw mill of Villard de Honnecourt. The caption reads in translation: "How to make a saw operate itself." (Reprinted by permission of Harvard University Press, publisher of Abbott Payson Usher's* A History of Mechanical Inventions. *This figure also appears on p. 55 of this volume.)*

Library of Congress Cataloging in Publication Data

Main entry under title:

Science and technology in medieval society.

(Annals of the New York Academy of Sciences, ISSN 0077–8923; v. 441)
Bibliography: p.
Includes index.
1. Science, Medieval—Addresses, essays, lectures. 2. Architecture, Medieval—Addresses, essays, lectures.
I. Long, Pamela O. II. New York Academy of Sciences. III. Series.
Q11.N5 vol. 441 500 s 84-27320
[Q124.97] [509'.02]
ISBN 0-89766-276-8
ISBN 0-89766-277-6 (pbk.)

CCP
Printed in the United States of America
ISBN 0-89766-276-8 (Cloth)
ISBN 0-89766-277-6 (Paper)
ISSN 0077-8923

ANNALS OF THE NEW YORK ACADEMY OF SCIENCES

Volume 441

April 19, 1985

SCIENCE AND TECHNOLOGY IN MEDIEVAL SOCIETY

Editor
Pamela O. Long

CONTENTS

Preface. *By* Pamela O. Long vii

Introduction. *By* Pamela O. Long 1

Rationality, Tradition, and the Scientific Outlook: Reflections on Max Weber and the Middle Ages. *By* Brian Stock 7

The Relevance of the Middle Ages to the History of Science and Technology. *By* Nicholas H. Steneck 21

The Mountain Men of the Casentino during the Late Middle Ages. *By* John Muendel 29

Artisans, Drudges, and the Problem of Gender in Pre-Industrial France. *By* Daryl M. Hafter 71

Where Roof Meets Wall: Structural Innovations and Hammer-Beam Antecedents, 1150–1250. *By* Lynn T. Courtenay 89

High Gothic Structural Development: The Pinnacles of Reims Cathedral. *By* Robert Mark and Huang Yun-sheng 125

The Function of Mechanical Devices in Medieval Islamic Society. *By* George Saliba 141

Fevers, Poisons, and Apostemes: Authority and Experience in Montpellier Plague Treatises. *By* Melissa P. Chase 153

Science for Undergraduates in Medieval Universities. *By* Edith Dudley Sylla 171

"Impious Men": Twelfth-Century Attempts to Apply Dialectic to the World of Nature. *By* Tina Stiefel 187

Optics through the Eyes of the Medieval Churchmen. *By* Samuel Devons 205

Preface

HALF OF THE papers in this volume were presented at a conference entitled "Science and Technology in Medieval Society," held on November 12, 1983, which was organized by the Medieval and Renaissance Studies Program at Barnard College and funded by the National Endowment for the Humanities. That meeting was the generative occasion for the present volume, which is not an exact record of its discourses, but a new collection of writings. The conference itself has a life of its own in the memories of those who attended, a tradition in the previous four conferences organized by the Medieval and Renaissance Studies program at Barnard, and a context of inspired undergraduate education and advanced research led by Professors Suzanne Wemple and Maristella Lorch.

PAMELA O. LONG

Introduction

PAMELA O. LONG

Department of History[a]
Barnard College
Columbia University
New York, New York 10027

WHILE THE PAPERS in this volume treat diverse topics in the history of medieval science and technology, they all reflect a relatively recent direction in the historiography of these disciplines, one concerned with the relationships between the content of scientific and technical knowledge and the society which both engendered that content and was, in turn, influenced by it. An emerging goal is to understand medieval science and technology on its own terms and as part of a specific social and cultural context. It is recognized that the practice of viewing such knowledge as if it had been developed within the framework of separate autonomous disciplines is derived more from the contingencies of modern historiography than from the realities of the medieval world. In spite of the methodological difficulties addressed by some of these papers, a study of the intrinsic relationship of the knowledge content to its social and cultural context promises to enhance our understanding both of the disciplines themselves and of medieval society.

The inclusion of both science and technology in the same volume may require some defense, since traditionally the two have been considered to occupy very different cultural spheres in the medieval period. Science has been viewed as part of a literate culture, whereas technology has been seen as belonging to a nonliterate culture. Yet to study science and technology as integral aspects of medieval society is to see them in closer relationship to one another than has usually been the custom; indeed, it is to give additional elaboration to Guy Beaujouan's thesis of the intimate relationship of theory and practice in the Middle Ages.[1]

The dichotomy between nonliterate "technologists" and literate natural philosophers has perhaps been overdrawn. Although no one denies that natural philosophers were literate, the paucity of medieval technical writings before the fourteenth century is as much as indication of the method of transmission of technical knowledge within a guild apprenticeship system as it is evidence for the level of literacy among skilled artisans. The few technical

[a] 1981–1982; present address: 2610 Cathedral Ave., N.W., Washington, D.C. 20008.

1

writings that we do have, such as the *Mappae Clavicula*, the writings of Eraclius, the eloquent treatise of Theophilus, and the comments of Villard de Honnecourt, testify to at least some medieval technical literacy. Moreover, it is significant that Theophilus presented not only detailed instructions for painting, glasswork, and metalwork, but also a defense on religious grounds for the openness of technical knowledge, against, one presumes, the prevailing habit of nondisclosure among the guilds.[2] Social systems and values profoundly affected the forms of transmission of both technical and theoretical knowledge, as well as its substantive content. And, of course, the reciprocal relationship of creation and transmission in the medieval period was quite different from what it is in our own age, in which technologies such as printing, television, and computers often make dissemination seem almost synonymous with technical and scientific authorship itself.

Our ability even to consider medieval science and technology in a social and cultural context is the result of the diligent study of texts and artifacts which, in the last sixty years, has created the foundations for these relatively new disciplines. It has been difficult to go beyond this basic work, perhaps because of the laborious and sustained efforts required to establish the textual and material bases of the fields, perhaps also because science and technology in the modern world have often been viewed as autonomous entities with their own imperatives, independent of their social and cultural environment. Nevertheless, until recently, the historiography of medieval technology has suffered somewhat less from tunnel vision than its counterpart in science. In part, this is because of the influence of Lucien Febvre and Marc Bloch, both through their particular contributions and in their general approach, which included the ideal, pronounced by Febvre in his prospective for a history of technology, that "l'activité technique ne saurait s'isoler des autres activités humaines."[3] That ideal has been carried forward by the influential American historian of technology, Lynn White, Jr., whose *Medieval Technology and Social Change* was dedicated to the memory of Marc Bloch.[4]

The papers of this collection are diverse: some consist of detailed particular studies, whereas others are more concerned with methodological issues. Both kinds offer new approaches that promise to open further productive avenues of inquiry. Brian Stock makes a critical assessment of the importance of Max Weber's work for the study of medieval science and technology, suggesting that Weber's concerns are as relevant to the medieval as to the early modern period, and indicating as well that many of Weber's unwarranted presuppositions about the medieval period are still in some sense a part of the viewpoint of present-day historians. Stock has shifted the field of inquiry as it relates to the issue of "modernization" and the development of a "scientific outlook."

The relationships of interiority and exteriority, and of rationality and tradition (as opposed to modernization), he suggests may be fruitful areas of investigation for understanding change in the medieval world.

Also concerned with the problem of change, Nicholas H. Steneck points to the difficulty scholars have had in determining the precise relevance of medieval culture to the scientific revolution. The shift from universalist to particularist thinking that took place in the fourteenth century, and its distinction between the absolute power of God and his ordained power (under which he created this world governed by rational laws) seem crucially important. Yet, the explanatory power of these notions for radical change in scientific ideas is lessened by the long delay between the two. Other factors have always had to be called into play. Steneck suggests that the change in these other factors (e.g., "changing methodologies, world views, occupational patterns, educational programs, support structures, interest priorities") seen as part of a fundamental shift in values, is the scientific revolution, and should be investigated from this point of view. Both Stock and Steneck have changed the focus of discourse about scientific and technical change in ways that have important methodological consequences.

John Muendel argues for an empirical, cumulative, and especially, regional orientation for the study of the history of technology, and then demonstrates the fruitfulness of such an approach in his paper on the ironworkers of the Casentino region of Italy. His study is based on a thorough analysis of the account books of two ironworkers whose primary occupation was the manufacture and repair of tools. His conclusions are as far-reaching as they are, at times, surprising. He brings into question the supposed dichotomy between the sophisticated town and the backward countryside of medieval Italy, and modifies the view of this particular region as impoverished and miserable because of its tenuous connection to urban centers. He demonstrates a subsistence barter economy, but one based on considerable technological capacity. It is notable that the ironworkers in question were literate and that they used a sophisticated accounting method. The organization of the shop, its acquisition of materials, the details of its production, and the methods by which it distributed its products are all elucidated in this study, as also are aspects of agriculture, transportation, and mills. Indeed, this study brings to question the extent to which the much lamented lack of written sources for early technology is due to the failure to utilize existing documents in the archives of Europe.

Social and cultural values influence the particular technologies that a society uses and also the ways in which those technologies are organized. The reciprocal relationship between technology and values is often mediated in

the arena of work. In this regard, Daryl M. Hafter's paper has particular relevance. In it, she explores the relationships between various technologies, access
to them, the organization of work, and gender in the Old Regime. She suggests that the poorest classes, by being denied access to a number of technologies, lost their gender identification in a situation of common oppression and
hardship, alike for men and women. An added inference is that they also lost
their identification as persons in contradistinction to animals. Hafter broaches
the complicated question of the participation of women in medieval French
guilds, and analyzes the ways in which they participated in technologies that
were often controlled by those guilds. She discusses the extent to which this
access was affected by the growth of royal authority, and emphasizes the centrality of the cloth industry both to women's work in medieval France and
to the medieval French economy.

If cloth making was the most important industry of medieval society, the
construction of the Gothic cathedral was its most spectacular achievement.
Both activities usually took place in the complex economic and social context
of the medieval town, and drew their workers from the same labor pool; indeed, the cloth industry provided some of the surplus wealth that was necessary for the proliferation of monumental architecture in the twelfth and thirteenth centuries. Taken together, the papers of Lynn T. Courtenay, and of Robert
Mark and Huang Yun-sheng, greatly enlarge our specific understanding of
the experimental nature of that architecture. These studies, which concern
the development of hammer-beam construction and the pinnacle, respectively,
show how a detailed and comparative study of structures can illuminate not
only some of the technical problems faced by medieval carpenters and masons,
and their particular solutions, but also some of the specific relationships of
technology and aesthetics in the development of the Gothic style. Social and
cultural historians may also be surprised to realize the considerable contribution that each of these studies makes to understanding the transmission of
technical ideas. Much has been written on the relationship of Gothic aesthetics
to the metaphysics of illumination and the theology of incarnation. These
papers do not obviate this understanding, but instead, firmly ground it in the
context of technical experimentation and the exchange of ideas among carpenters and masons.

It is the lack of an extensive associated textual tradition that situates the
study of the Gothic cathedral almost exclusively on the investigation of the
buildings themselves. In the opposite circumstance, George Saliba treats a textual tradition (Arabic treatises on mechanical devices) for which the material
objects under discussion are often not extant. Yet, by his careful examination
of these texts, Saliba has been able to revise our understanding of the function

of these treatises and of the devices that are their subject. He has placed them in the tradition of the Aristotelian *Mechanics* and has shown that, in addition to the practical uses of some of the devices, they were considered to represent the actualization of principles that existed in potentiality. He finally emphasizes that the beauty of the illustrations and of the objects themselves did not diminish their functionality.

In a very different context, Melissa P. Chase has also scrutinized a textual tradition — plague treatises associated with the medical school at Montpellier — to gain a fuller understanding of the relationship between theory and practice. She shows how traditional theories of medicine were modified by the experience of the plague and by physicians' attempts to understand and treat the devastating epidemics of this disease. Using treatises from the first outbreak in 1348 to those of the mid-fifteenth century, she has been able to demonstrate a development of ideas about the etiology and taxonomy of the plague. She shows how experience rendered inadequate the traditional theory of disease based on the imbalance of the humors. By pointing out the increased importance of apostemes and poisons as explanatory devises, she explicates the theoretical and practical results of this development.

Edith Dudley Sylla treats the function and use of scientific texts within the institutional setting of the university. Her assessment of the scientific works encountered by undergraduates furthers our knowledge of medieval university education and of the scientific sophistication of the average student. In her attempt to understand the motivation of the medieval arts student in studying science and mathematics, Sylla emphasizes that the methods of the disciplines were similar, whether they concerned science, mathematics, philosophy, or theology. The mastery of the practice of teaching was one of the major goals of studying the texts; as important as the particular content of scientific and mathematical books was the established methodology of using and interpreting them. The identity of method (the scholastic) in both teaching and doing science blurred the distinction between pedagogy and research. Her discussion makes clear that understanding the uses of texts can be a necessary concomitant to understanding their contents.

Well before the development of scholasticism, in the early twelfth century, Tina Stiefel argues, new attitudes developed among a small group of writers. A new confidence in reason and in man's ability to discover rational laws opened up the world of nature as a legitimate area for scrutiny and investigation. Since the assumption that nature is susceptible to rational understanding does seem to be a necessary precondition for science, the study of confidence in reason, and of ideas about the uses of reason are crucially important to understanding the development of a "scientific outlook." Specifying some of the available classical sources, Stiefel emphasizes that this new confidence developed before the complete Aristotelian corpus had been translated into Latin. She also points to the charges of impiety that were leveled against these early "cosmologists," showing that their new attitude toward nature was both derived from a modified notion of man's relation to God, and came into conflict with the established view of that relation.

Yet by the thirteenth century science and theology had become subjects of reciprocal reinforcement, as Samuel Devons illustrates in his paper on optics and theology. To investigate light and vision, the eye, refraction and reflexion, and to study the rainbow were acts of piety that brought the investigator closer to divine illumination. Such activities were no longer considered an affront to the belief in God's power to produce miracles. It is significant that many of the investigators of *perspectiva* were also leading figures of the church or of the orders. It is also notable that their discussions include references to experience and experiment, although the extent to which they employed experimental methods remains unclear.

The papers in this volume demonstrate the value of studying medieval technology and science as part of a social and cultural context: not only does this method of study produce a more coherent view of medieval life and thought, it allows a greater understanding of the content of medieval scientific and technical knowledge. Many of the scholars herein have also contributed to the development of methodology by using innovative approaches to their sources. And it is likely that the results of such research will be increasingly fruitful. For as the value-laden nature of modern scientific and technical disciplines becomes increasingly understood and studied in detail, it is likely that we will more readily find methods and sources for understanding the social and cultural contexts of their counterparts in the Middle Ages.

NOTES AND REFERENCES

1. *L'interdépendance entre la science scolastique et les techniques utilitaires*, Les Conferences du Palais de la Decouverte, Ser. D, No. 46 (Paris, 1957); and "Réflexions sur les rapports entre théorie et pratique au moyen âge," in John Emery Murdoch and Edith Dudley Sylla, eds., *The Cultural Context of Medieval Learning* (Dordrecht and Boston: D. Reidel Publishing Co., 1975).

2. *De diversis artibus*, translated and edited by C.R. Dodwell (London: Thomas Nelson and Sons, Ltd., 1961), pp. 2–4 and 61–62.

3. See particularly the November 1935 issue of the *Annales d'histoire économique et sociale*, which was dedicated to the subject "Les techniques, l'histoire et la vie." For Febvre's comment, see "Réflexions sur l'histoire des techniques," p. 532.

4. (Oxford: Oxford University Press, 1962).

Rationality, Tradition, and the Scientific Outlook: Reflections on Max Weber and the Middle Ages

BRIAN STOCK

Pontifical Institute of Mediaeval Studies
59 Queen's Park Crescent
Toronto, Ontario, Canada M5S 2C4

I BEGIN WITH a generalization with which, I trust, most historians would agree. During the past fifty years progress has been made in our understanding of virtually every aspect of medieval science and technology; yet we have at present no commonly accepted model for interpreting these developments within the broader environment of medieval society and culture. What we have in fact are bits and pieces of various models: these are rather loosely tied together by a number of assumptions that are too rarely examined in depth.

If we are to find a way out of this difficulty, we must attempt to address two problems at once. One concerns inherited interpretations, which, with Thomas Kuhn, we may call the normal science of medieval science. The other involves the broader social and intellectual context of premodern achievement as a whole. Why, we must ask, did medieval science and technology develop in a manner uncharacteristic of the ancient world, or, for that matter, of any previously known society?

These are weighty issues, and I do not pretend that I can review them adequately in a brief presentation. Yet, by focusing our attention on a few central questions which have been raised in one form or another again and again, I hope at least to provide a starting-point for some common reflection among historians, both of science and of culture, who too often view their contributions as isolated endeavors rather than as elements in an integrated mosaic of ideas.

By stating my theme in somewhat negative terms, I do not mean to suggest that there have been no attempts to come to grips with larger issues in recent years. Quite the contrary: the malaise to which I refer has been felt by a number of students, most notably by Lynn White Jr., and the lack of general hypotheses is all the more noticeable in view of the important advances that have been made since the publication of Haskins' *Studies in the History of Mediaeval Science* in 1924. The collective history edited by David Lindberg in 1979 highlighted some of the principal topics of interest: the textual transmission of Greek and Arabic learning, the appreciation of the institutional and philosophical contexts, and the interdependence of such key fields as mathematics, statics, dynamics, optics, astronomy, medicine, and natural history. Larger questions have also been raised concerning such matters as science and magic, techniques and social organization, and, in Alexander Murray's substantial study,[1] the relationship between reason and other aspects of intellectual development.

But posing questions and providing answers are two different matters. The growing presence of larger concerns in the study of medieval science is significant. Yet their influence on general modes of interpretation has been slight. It is an unfortunate fact, although not often enough remarked, that, as soon as one turns from specific disciplines and devices to more general issues, the historiographical picture begins to blur. By and large, we still view the field as a set of isolated success stories, with both the benefits and the limitations that a unilinear, evolutionary, and internalist tradition implies. If our understanding is to be anything more than that, we cannot be satisfied with vague expressions of interest: we must return to basic points of departure and think certain problems through anew.

Those who work on medieval science tend to trace the origins of serious professional interest in the subject to such influential figures as Duhem, Sarton, and Thorndike. Students of technology are less conscious of historical alignments, partly because of the fragmented nature of technical invention and partly because researchers in the field have less persistently sought the intellectual respectability which, rightly or wrongly, is conferred by association with a theoretical science. But in recent years the history of technology has also begun to lean towards what Derrida has aptly termed logocentrism. However, if we wish to examine the cultural roots of the scientific outlook, in which medieval achievement is deeply implicated, we must acknowledge that the beginnings of serious reflection lie in a completely different area. It was in fact Max Weber who first focused our attention on the potential links between science, technology, rational action, and economic or social circumstances. Whether we like it or not, we as medievalists are no less heirs to the Weberian tradition than the early modern historians to whom he more particularly directed his studies. If we are to straighten out our thinking about the cultural background of science, we must first straighten out our thoughts about Weber.

Later I shall draw attention to a number of positive and enduring features of Weber's mind; and, because there is a critical tone in what immediately follows, I should like to stress at the outset that I am in no sense an anti-Weberian. Yet, as regards medieval science and technology, it is necessary to begin any reconsideration by pointing out three unproductive features of his outlook. These concern his understanding of medieval thought, his overuse of the ideas of tradition and modernity, and his notion of ideal types as functional intermediaries between social and intellectual change.

In order to put my remarks in an acceptable context, I offer by way of introduction a few observations on the changing views of Weber's sociology of religion over the past three generations.

Broadly speaking, the evolution of Weber's views can be divided into three phases. In 1904 and 1905 he published his well-known essay, *Die protestantische Ethik und der Geist des Kapitalismus*, and he spent roughly the following five years defending his thesis on the connection between Calvinism and western capitalism. Between 1911 and 1913 a second phase was begun, during which Weber compiled the section on the sociology of religion in *Wirtschaft und Gesellschaft*, incorporating the important distinction between asceticism and mysticism from his reply to Ernst Troeltsch at the first meeting of the Deutsche Gesellschaft für Soziologie at Frankfurt in 1910.[2] The third phase consisted of a set of revisions of earlier material which appeared as articles in the *Archiv für Sozialwissenschaft und Sozialpolitik* between 1914 and 1917 and were later collected, along with a somewhat revised version of *The Protestant Ethic*, in the three-volume *Gesammelte Aufsätze zur Religionssoziologie*, the first volume of which Weber was himself editing shortly before his death in 1920.[3]

Early studies of Weber, particularly in the English-speaking world, directed their attention chiefly to the problems raised by the Protestant ethic thesis, which has by now been amply, perhaps too amply, discussed at both an historical and a sociological level.[4] Yet the emphasis, however merited, which was placed on this issue inevitably left other aspects of his complex intellectual development in neglect. Early students of Weber showed little appreciation of how his family's evangelical background had helped to shape his interests,[5] nor of how his ideas on power, authority, and legitimation grew naturally out of his perception of the German political situation before World War I.[6] Concentration on a single work within a large, unfinished, and somewhat disorganized corpus tended as well to give a monolithic character to his thinking: it isolated him artificially from Marx and Nietzsche, as well as from his friends like Tönnies, Simmel, and Troeltsch, with whom he discussed and debated his ideas. Only with the appearance of mature intellectual biography, and along with it a better understanding of the chronological growth of his ideas, has the full range of Weber's interests become known. The shift in emphasis is particularly noticeable in the interpretation of his sociology of religion. The comparative as opposed to the purely evolutionary aspect of his investigations has been made clear. Research has been diverted from *The Prot-estant Ethic*, and to a lesser extent from the methodological introduction to *Economy and Society*, to the final statement of his views in *The Economic Ethics of World Religions*.[7] Above all, the earlier focus on theory and meth-odology to the neglect of historical considerations has been replaced by a more balanced view, which sees one of the recurring issues in his work as the process of rationalization in western society. Rationality, in this sense, is no longer analyzed in isolation as the promoter of one economic outlook, but is rather looked upon, as Weber intended, "within the totality of cultural development."[8]

The research trends that I have briefly summarized have by and large taken shape within German scholarship since World War II. They allow us as never before to distinguish between Weber's own thought and the mass of interpre-tive material that inevitably springs up around a seminal thinker; they also allow us to isolate certain imperfections in Weber's historical and methodo-logical constructions, the blame for which can no longer be laid on the door-step of his North American enthusiasts.

One of these, as noted, results from Weber's rather inadequate education in medieval philosophy and theology. Even those who oppose Weber's interpre-tation of Reformation thought acknowledge that he made useful distinctions among the world views of the Lutherans, Calvinists, Pietists, Methodists, and Baptists. But he was in the habit of lumping all pre-Reformation theology together in a single, synthetic picture to which he referred on more than one occasion simply as "Catholic thought." Weber seems to have made a number of questionable assumptions about the homogeneity of other aspects of me-dieval cultural life. He took the dogmatic affirmations of councils, the papacy, and official thinkers as the norm, not considering the wide diversity of views which persisted throughout the Middle Ages, even among Christians. He also assumed a far closer correspondence between the ideological and institutional aspects of the Christian life than existed in reality. Worst of all, he concen-trated his interests within too narrow a time frame. Like Marx, with whom in this respect he had much in common, he saw a single great period of change at the end of the Middle Ages with the breakup of feudalism and the rise of urban society: he paid little attention to the centuries that intervened between the Carolingian dynasty and the fourteenth century, a period whose richness and diversity has increasingly occupied the attention of historians of the me-dieval and early modern periods. His grasp of medieval mentalities and of forms of thought was consequently highly limited. I am not urging the in-defensible position that there are no differences between Catholic and Protes-tant theologies or that there was no parting of the ways between medieval and postmedieval philosophy. However, it is a weak intellectual method to

look upon one group of thinkers as operating outside institutional boundaries and to look upon another as being entirely constrained by them. There
was more conformity in Reformation thought than Weber admitted, and more
productive variety in the Middle Ages.

The second weakness in the medieval aspects of Weber's sociology of religion concerns the much discussed relationship between tradition and modernity. Weber's model of rationalization brilliantly accounted for the advent and
progress of modernity; rationalization in this sense is virtually equatable with
modernization. But he failed to account adequately for the continuity and
evolution of tradition, or for that matter to define what he meant by tradition
at all. In Weber, Edward Shils notes, "the persistence of past practices and
arrangements is not taken up as something to be explained."[9] And what is
not explained is assumed not to have changed. Transferred to the historical
dimension and put rather crudely, this means that, broadly speaking, the
Middle Ages is identified with tradition and the Reformation with modernity: the Middle Ages is therefore static, the post-medieval period dynamic.
Three generations of research have produced a more nuanced picture, in which
"the revolt of the medievalists," as Wallace Ferguson termed it, has modified
many of the distortions of earlier Renaissance and Reformation scholarship.
Yet the new perspectives have had much more influence on monographs,
journal articles, and specialist literature than on more general historical theses.
Indeed, a version of the early modern hypothesis — which effectively com-
presses change into a brief transitional period between the Middle Ages and
the Renaissance — still dominates much investigation into the borderland area
between science, other types of causality, and society. No better example of
this approach can be cited than Keith Thomas's admirable *Religion and the
Decline of Magic*, in which the remarkable discussions of astrology, prophecy,
witchcraft, and other occult beliefs are virtually all played out against the
backdrop of what is called "the magic of the medieval church." A more recent
book which perpetuates the same viewpoint is Timothy Reiss's wide-ranging
study, *The Discourse of Modernism*.

Once again I should like to make my position clear: I am not arguing that
there was no transition from "tradition" to "modernity," however we call it,
nor that the case of historical understanding will be much advanced if we
try, as some wrong-headed medievalists have tried, to push the Reformation
back into the Middle Ages. Yet, by applying the tradition/modernity dichotomy
to areas of life and thought in which it does not belong, we inevitably finish
in a position of ideological polarization: by the way in which we pose the
problem we rule out asking important questions, and, as a consequence, we
oversimplify historical development generally. In point of fact, since Weber

made his woeful prediction about the imminent extinction of tradition, modernization has not been such an ineluctable force as he envisaged, and, if one
looks at the cases of Japan, Israel, or Iran, tradition and radical change have
not always been incompatible. In medieval history, too, the further back one
ventures, the more significant and multifaceted is the role of the many forces
we facilely group under the name of tradition. If one wishes to achieve a valid
overview of how science and technology interpenetrated medieval society between the eleventh and the thirteenth centuries, one must begin by broadening
the notion of tradition at least far enough for it to become a workable analytical tool. For, if medieval society did not change through a link between rationality and modernization, as, it is arguable, did preindustrial England or
Germany, then it must in part at least have changed through unexplored links
between rationality and tradition. As curious as it may seem to a generation
of historians reared on the notion of unidirectional progress, medieval rationality and rationalization, to use Weber's favorite terms, were actually
byproducts of the intensification of tradition.

The third major imperfection in Weber's sociology of religion as it applies
to medieval research is a methodology which for its time and place is one
of its chief strengths, namely the employment of ideal types. By their use,
cultural change is viewed as, if not determined by, certainly acting in close
harmony with economic and social transformations. Studies attempting to
explain relations between culture, society, and economy have long operated
under the assumptions of what is in fact a tripartite construction of reality,
in which the consideration of religion or the intellectual life is subordinated
to the facts of material life and social organization. This model—if one may
now comfortably speak of it in the past tense—had the virtue of reflecting
the Marxist concern with superstructure as well as a type of functionalism
in which, Clifford Geertz notes, the rituals, symbols, and ideas interdepen-
dent with behavior were taken to be supportive of the social materials out
of which they were constructed. The historical counterpart of this approach
leads us once again to the worst features of what I have called the early modern
hypothesis. Presented in its most general form, the argument asserts that, because we associate the scientific revolution with the sixteenth and seventeenth
centuries, we must seek from within that time frame all the other components
of the modernization process, relegating to the Middle Ages, as to all earlier
epochs, mere anticipations of modernity. As a consequence of this hypothesis, particularly in textbook versions of the past, most contemporary interpretations of modernization are emphatically postmedieval in orientation. In fact,
one of the unfortunate byproducts of this type of functionalism is to legiti-
mize ignorance of the past—why study remoter periods of history, after all,

if they are only marginal in importance — and to strengthen the bias of the traditional/modern axis of inquiry. At bottom this is a form of educated anti-intellectualism. And, to the degree that medievalists have seen their task as pushing back the boundaries of modernity, they have become unwitting accomplices in a conspiracy to assign their efforts to the realm of historical irrelevance.

The answer to functionalism is not nonfunctionalism, that is, the belief, maintained by historians of the nuts-and-bolts variety, that nothing is provably related to anything else, and that, as a result, history is just a story in which one event simply follows another. Also, in attempting to redress the functionalist bias, we must take care not to forget the lessons that recent theory and practice have taught us. From linguistic philosophy we have learned that all interrelations can be questioned insofar as they are statements or propositions, and that it is difficult to separate events as pure occurences from the language in which they are described. The lessons of social history are even more important: the concern for getting straight the facts of material life, the relevance of counting all countable information, no matter how marginal, and the rejection of the naive history of ideas, which artificially removes individuals from their times, places, and settings, and only takes account of abstract links between concepts.

Yet we must also distinguish in history between the objective and the subjective. What men regard as life's physical or social necessities is often shaped by pre-existing notions that are not reducible to elements present at a given moment in the environment. Weber based his interpretation of capitalistic development on the notion of meaningful social action, where the stress on meaning relates to the subjective orientation of the individual, not to the mere simultaneity of occurrences in time. We may put this another way by distinguishing as does Saussure between association and syntagm or as does Jakobson between metaphor and metonymy. This means that historical events, like words in a language, may occur in a sequence of expression without necessarily being related in a meaningful way. It is not only temporality which determines cultural sense and logic but also the inner significance that individuals and groups assign to events, whether these take place in the real world, or, as is often the case, they are creations of the imagination. As cultural meaning transcends the individual and his or her world, so it transcends the historical slice of time out of which functionalism derives its basic relationship. What our present understanding amounts to is this: certain causal factors in any complex historical situation are interrelated through the same time frame, as well as by function, while others are not. Functionalism is just one interpretive field among others, and its correlations must be established by empirical investigation, not derived from *a priori* theory. If we turn these criticisms toward the science and technology of the Middle Ages, it is evident that emphasis has been placed on some issues to the neglect of others. As a byproduct of crude Weberianism, a great deal of attention has been paid to the late, almost premodern phases of development. But we have only begun to understand the cultural and religious forces which properly belong to the earlier period.

To summarize to this point: the reasons why the Middle Ages contributed decisively to the advent of modernity, that is, to the conjunction of science, technology, economics, and mental attitudes that we associate with the modern world, cannot be found by means of a number of tried historiographical approaches, even though each has added its share of insights to the process of modernization as Weber understood it. These methods include: the internalist history of science and technology, the unmodified Weberian sociology of religion, the mere pushing back of the boundaries of modernity into the Middle Ages, or any sort of functionalism, which reduces cultural change to a superstructure of economic and social transformations. However, if this criticism of earlier schema is accepted, then it follows that the essential changes of the Middle Ages may not have taken place in science or technology as such, but rather in more subtle mutations in the cultural and religious environment, which not only prepared the way for the reception of modern attitudes, but, more importantly, began the debate, which continues down to our own day in the less developed world, on whether this sort of rationalization was a good thing.

The essential task, it follows, is not just another defense of the Middle Ages, or worse, a defense of medievalism, but the patient deciphering of the issues, both for and against, which provided the background for change and its associated conflicts. By way of conclusion, let me provide two examples of the sort of problem I have in mind.

One of the fundamental questions which an inquiry into social change must pose is the relationship between thought and action. This issue takes precedence over whether, within the modernization process, we consider action rational in itself, or whether we judge it to have fulfilled allegedly rational goals. Before such matters can be discussed, we must know what mental equipment existed for allowing individuals to look upon transformations in the outside world as resulting from human will, motivation, or intellectual design.

The answer to this question brings us face to face with the problem of interiority and exteriority as it developed between the Carolingian age and the high Middle Ages. At the most basic level this meant nothing more or less than an exploration of the mind as an inner world; it was, so to speak, a rediscovery of consciousness and the unconscious. First revealed as a sentiment, a feeling, or an attitude, the notion of interiority progressively became conceptualized: one finds it reappearing in different areas of thought under such rubrics as intention, abstraction, or subjectivity, as well as in the theological ideas of penance and casuistry. A link was found between sin, ethics, and broader human aims, and, together with the nascent sensations of guilt and shame, entered both pastoral treatises and Old French romances. At a somewhat more rarefied intellectual level, a debate was carried on over alienation, that is, in medieval terms, the manner in which man's image, viewed as an inner reality, differed from God's design after the Fall, and the manner in which man could through active effort repair some of the damage. Psychological reflections, often based on analogy with the Trinity, became fashionable, and utopian thinking became less concerned with finding a lost garden or paradise than with achieving a real or imaginary goal that did justice to reform as an inner and future ideal. Still another facet of the same complex of ideas was the rise of visionary and prophetic literature, both in Latin and in the vernaculars, which explored hitherto uncharted regions of memory, nostalgia, and unfulfilled promise. These changes were not limited to the literary world; through allegory, which acted as a collecting point for the new sentiments, the notion of an inner life or sense became part of sculpture and manuscript painting. And in the predecessors of St. Francis — the wandering preachers, monastic reformers, and less orthodox figures who crowd the stage of twelfth-century religious life — interiority and exteriority gradually fused into the notion of *imitatio*.

The chief mechanism by which the interior and exterior aspects of life were brought into harmony or perceived to be in disharmony was the literary, theological, or visually reproduced text. From the outset, inner consciousness was brought about by the reading, study, and meditation upon texts. In every field, the rediscovery of the spiritual reality of the mind was a sort of stepchild to a textual experience which began in this world with words, prayer, recitation, and liturgy, and gradually worked its way into an entirely abstract, synchronic presence, in which the written word, or, as it was preferred, its spoken equivalent, appeared to be only an image of a deeper truth. Solitude was essential; the reader had to be alone with his text. One had to leave group for individual hermeneutics; the exterior had to give birth to the interior text in the mind. Meditation and solitude thereby became the preconditions of man's realization of the existence of his inner self, and, very often, his only form of verification: the notion that, in contrast to the random, heterogeneous, and often meaningless record of external events, there was a potential pattern for his life and actions which could be established through the continuity of mind or spirit. One witnesses the birth of the idea of a spiritual direction for history based on the inner continuity of ideas, traditions, and institutions, and, in Joachim, the first intimations of a Hegelian type of dialectic, which foresees that different "spirits" dominate the revolutionary passage from one historical period to another. At a more mundane level, the production of speech acts and the swing from inner to outer linguistic statement appeared in many areas of life and thought. The period's most symptomatic figure is not perhaps the great thinker, who, like Abelard, teaches a small circle of devotees, but rather the preacher, for whom there is a constant tension between word and deed, and who conveys this tension for the first time to the masses.

One issue, then, as I see it, is that of thought and action, of models and realities, or, as I have called it, of interiority and exteriority. A second major question concerns rationality and rationalization, or, to put the matter once again in a medieval context, the interrelated questions of rationality, tradition, and design.

For Weber, of course, the problem of rationality came first, and was linked to his divisions of meaningful action, whence his pessimistic view, shared by Nietzsche, that "occidental rationality" was only achieved at the expense of a new type of enslavement, which, ironically, was brought about by the process of rationalization itself. Students of Weber, by and large, have looked upon rationalization as a force working on man from the outside, and medievalists from Haskins on have devoted much attention to the history of administration and bureaucracy, through which, Weber argued, formal rationality is legitimized and made part of everyday routine. Recent studies of Weber's notions of rationality, which, for better or worse, he never systematically sorted out, have taken a different turn. If we take the current view that Weber had in mind essentially four types of rationality, namely the practical, the theoretical, the substantive, and the formal,[10] it can be shown that he devoted more attention than was previously thought to the notion of substantive rationality, that is, to a type of rationality that directly orders action into patterns, not, as in practical rationality, by a simple means-end calculation (*Zweckrationalität*) but in relation to a past, present, or future construction of values (*Wertrationalität*). In an important but often overlooked note in *Economy and Society*, Weber himself distinguished between formal rationality, by which he meant the potential and applied degree of quantitative calculation in a means-end relationship, and substantive rationality, under which rubric he grouped all other economically ordered social action oriented by some past, present, or future criterion of ultimate values, regardless of how these are defined.[11] To bring the question into line with contemporary debate, we need only add a note of relativism. Substantive rationalities are not always economic in origin or consequence, as Weber seemed to suggest, and, as it turns out, not always limited to "western" societies, as the research of Maurice Godelier, Bryan Wilson, and Claude Lévi-Strauss has shown.

I referred above to a link between rationality and tradition, which, I suggested, was a more promising line of inquiry than the mere search for the roots of modernity. In one sense, it does not matter whether we call this phenomenon traditionalization or modernization, since the results were precisely the same, namely a changed social and intellectual environment. But it does matter if we wish to offer a coherent explanation why such changes took place. In point of fact, true modernizers, in Weber's terms, were in the minority in the two centuries before 1200. The most frequently employed channel for alterations in *mentalité* lay between rationality and tradition; the link gains greater force if we take into consideration not just the objective mutations but also the manner in which individuals explained them to themselves. We witness what has been called "substantive traditionality,"[12] that is, a type of tradition which was not just a pattern of inherited thought or conduct but which represented a new type of activity, a conscious attempt to get beneath the encrustations of history, and, as a consequence, to order behavior according to a traditional meaning that had been reflected upon and measured against known standards in the past. The development begins as early as Peter Damian, whose many statements on tradition have never been systematically studied, and continues into the twelfth century in such authors as Humbert of Silva Candida, Honorius Augustodunensis, and Gerhoh of Reichersberg. These are not theorists in the normal sense, nor do they simply oppose rationality with symbolism. Although they occasionally utilize theory, they see as their primary task the corroboration or correction of contemporary practices in the Church in relation to the norms laid down by Jesus, the apostles, and the fathers. These prescriptions, when they were set in order through the nascent disciplines of law and church history, appeared to them to be both traditional and reasonable; therefore, rationality dictated that all future practice should be based upon them. However, in order to reform practice, one first had to have an accurate idea of what was originally thought and done. This required the elimination of elements inconsistent with the original design. I deliberately employ the metaphors of literature because the activity was much more like literary history than rationalization in the contemporary economic and social sense. But the consequences for what we know as rationality were far-reaching.

Time does not permit me to discuss a range of other problems touched upon by Weber and bearing upon the subject, such as work and leisure, voluntary association, and the uses of knowledge, and I can only hope that such a brief and all too general survey has persuaded you that I am correct in laying the emphasis where I have. The history of science and of technology in the Middle Ages must of course continue to be written in the traditional way: the patient scholarly work of editing texts, providing fundamental commentaries, and sorting out the individual and institutional contexts. But one must also step back from the field from time to time and ask where we are going. If we are to understand, not only the individual stages in the process, but why the debate on modernization began when it did and gradually assumed its characteristically western form, then we must come to grips with the value system of the Middle Ages, and above all let it speak for itself. Max Weber had the privileged view of a German neo-Kantian at the outbreak of World War I. It was possible for him to criticize rationalization while still making human reason his criterion for all value systems. Four generations later it would be naive indeed for anyone to speak of our century in terms of a unilinear growth in rationality. Rationalities have been achieved in some areas at the expense of irrationalities in others; at times, rationalities and irrationalities have succeeded one another in no particular pattern. Perhaps, as I have suggested, the Middle Ages was more like the latter alternative than has been realized: rationality was not only a positive force in history, shaping both men and ideas, but also an image, divinely or demonically inspired, which man kept before himself in order to disguise the ultimately less rational sources of his desires and ambitions.

NOTES AND REFERENCES

1. *Reason and Society in the Middle Ages* (Oxford: Clarendon Press, 1978).
2. "Diskussionsrede zu E. Troeltschs Vortrag über 'Das stoisch-christliche Naturrecht,' " in Max Weber, *Gesammelte Aufsätze zur Soziologie und Sozialpolitik* (Tübingen: J.C.B. Mohr [Paul Siebeck], 1924), pp. 462–470.
3. For the dates of the various compositions in the volume, see D. Käsler, "Max-Weber-Bibliographie," *Kölner Zeitschrift für Soziologie und Sozialpsychologie* 27:703–730 (1975).
4. The original criticism and replies are reprinted in J. Winckelmann, ed., *Max Weber. Die protestantische Ethik II. Kritiken und Anti-kritiken*, 2nd edit. (Hamburg: Siebenstern Taschenbuch, 1972), pp. 11–345. The subsequent literature is immense; for a full review up to the time of its publication, see C. Seyfarth and G. Schmidt, eds., *Max Weber Bibliographie. Eine Dokumentation der Sekundärliteratur* (Stuttgart: Enke, 1977).
5. See A. Mitzman, *The Iron Cage. An Historical Interpretation of Max Weber* (New York: Knopf, 1970), pp. 15–38.
6. In general, see W. Mommsen, *Max Weber und die deutsche Politik, 1890–1920*, 2nd edit. (Tübingen: J.C.B. Mohr [Paul Siebeck], 1974).
7. See the much discussed essay of F.H. Tenbruck, "Das Werk Max Webers," *Kölner Zeitschrift für Soziologie und Sozialpsychologie* 27: 663–702 (1975); partial English trans., "The Problem of Thematic Unity in the Works of Max Weber," *British Journal of Sociology* 31: 313–351 (1980).

8. *The Protestant Ethic and the Spirit of Capitalism*, trans. T. Parsons (New York: Charles Scribner's Sons, 1930), p. 284, n. 119.

9. *Tradition* (Chicago: University of Chicago Press, 1981), p. 8.

10. S. Kalberg, "Max Weber's Types of Rationality: Cornerstones for the Analysis of Rationalization Processes in History," *American Journal of Sociology* 85: 1145-1179 (1980). For a similar view, see D.N. Levine, "Rationality and Freedom: Weber and Beyond," *Sociological Inquiry* 51: 12-15 (1981). For a recent review of the issues, see W. Schluchter, "The Paradox of Rationalization: On the Relation of Ethics and the World," in G. Roth and W. Schluchter, *Max Weber's Vision of History* (Berkeley, Los Angeles, and London: University of California Press, 1979), pp. 14-15 *et passim* and, more generally, C. Seyfarth and W.M. Sprondel, eds., *Max Weber und die Rationalisierung sozialen Handelns* (Stuttgart: Enke, 1981).

11. Ch. 2.9, section 3, in *Wirtschaft und Gesellschaft. Grundriss der verstehenden Soziologie*, 5th edit., J. Winckelmann, ed. (Tübingen: J.C.B. Mohr [Paul Siebeck], 1972), p. 45.

12. E. Shils,[9] pp. 21, 287.

The Relevance of the Middle Ages to the History of Science and Technology

NICHOLAS H. STENECK

Department of History
University of Michigan
Ann Arbor, Michigan 48109

PRIOR TO THE LATE nineteenth century, few scholars assigned much importance to the Middle Ages when discussing the development of modern science and technology. Most assumed that the medieval period was best seen as a backward age whose darkness helped set out the brilliance of the Scientific Revolution and Enlightenment. Beginning in the first decade of this century, Pierre Duhem tried to change this view by presenting evidence of important scientific discoveries that were supposedly made in the fourteenth century by Parisian and English scholastics. By the late 1920s, Charles Homer Haskins drew back the curtains even farther with his speculations about the importance of the renaissance of the twelfth century.[1] With the renewed interest in medieval science and technology that followed, the way was opened for a critical reassessment of the relevance of the Middle Ages to the development of our modern scientific-technological world.

However, the process of reassessment has yet to produce any consensus on the relevance of the medieval period to the scientific and technological revolutions of western society. Duhem's continuity thesis is now seen as much too optimistic. Fourteenth-century thinkers criticized elements of the Aristotelian world view, but they did not abandon it. Their teachings on astronomy did not anticipate Copernicus's contributions to astronomy, nor did they discover inertia or analytical geometry. Save for a few possible minor exceptions from astronomy, optics, and mathematics, it is difficult to point to any speculations about nature maintained in the Middle Ages that were not later overturned during the course of the Scientific Revolution.

When the notion of direct intellectual ties became untenable, scholars began looking for evidence of indirect relevance, seizing next on methodology. John R. Randall and Alistair Crombie tried to demonstrate that Galileo's scientific methods had been anticipated by earlier Italians and before them scholastics of the high Middle Ages, such as Robert Grosseteste and Roger Bacon.[2] But again, the significance of the connection faded upon closer examination. Galileo scholars have since demonstrated that the roots of Galileo's scientific methods cannot easily be traced to a single tradition, such as the hypothetico-deductive method advanced by earlier scholastics, and the origins of the scholastic methods in question are themselves traceable to fundamental Aristotelian teachings.

With the rejection of both methodological and intellectual ties, historians turned to less direct connections to establish relevance for the medieval period. One way this was done was by giving a new twist to a thesis first advanced by Duhem. Duhem had argued that the ban on the teaching of some aspects of Aristotelian science in the well-known Condemnation of 1277 forced later scholastics to adopt new ideas to replace those that had been condemned. So it was, Duhem argued, that fourteenth-century scholastics anticipated Copernican astronomy when responding to the prohibition against teaching that "God could not move the world with rectilinear motion."[3] Subsequent scholars, who, as mentioned, for the most part have rejected this aspect of Duhem's work, have nonetheless continued to study the response to the Condemnation of 1277 for evidence of more subtle transitions that could have paved the way for later scientific developments.

Of particular promise in this regard seems to be an intellectual distinction fourteenth-century scholastics used to get around the rigid limitations posed by the Condemnation of 1277, the distinction between God's two powers — absolute and ordained.[4] The theologians who promulgated the Condemnation of 1277 objected to limitations that so-called radical thinkers were placing on God, *e.g.* that God could not move the world with rectilinear motion. Fourteenth-century scholastics circumvented this objection by agreeing that in accordance with absolute power God could do anything, but that under the agreement of ordained power God had produced the specific world that exists, which world could not be moved in a straight line, abhorred vacuums, and was centered around a stationary earth. In this way they were able to discuss in an imaginary way things that might have or could have been, while keeping to the orthodoxy of the accepted Aristotelian world view. Some scholars believe that this compromise represents a significant step toward the Scientific Revolution. But what was the nature of that step?

Amos Funkenstein has argued that significant advances were made in the course of "absolute-power" (*de potentia absoluta*) arguments. Whether or not fourteenth-century thinkers ultimately accepted such arguments is not what really matters, according to Funkenstein. Rather, the significance of the *de potentia absoluta* argument is that in bringing the improbable into clear view and in subjecting impossibility to a critical analysis, it prepared the way for the ultimate acceptance of the improbable and impossible. He suggests, therefore, that we ought to pay close attention to the terms in which a theory defines "improbabilities" and, still more important, "impossibilities." The more precise the argument the likelier it is to be a candidate for future revisions. Once the impermissible assumption is spelled out with some of its consequences, it is but a matter of time and circumstances (. . .) until the truly radical alternative is reconsidered.[5]

There is, according to this view, still some merit to the hypothetical discussions of the fourteenth century, even if that merit does not involve directly the people advancing them. By mapping out the ground on which the battle would eventually be fought, fourteenth-century thinkers were participating, however tangentially, in the Scientific Revolution that was to follow.

Heiko Oberman and Gordon Leff believe that it is the *absoluta-ordinata* distinction itself that is of most importance. Both argue that the individualizing, as opposed to universalizing, thrust of the two-power distinction, combined with the fact that science operates almost exclusively in the ordained world and without the aid of revealed truths, left the study of nature in a state of relative freedom — from theology and metaphysics — in the fourteenth century and in need of developing methods of its own for discovering truth. The results, again for both Leff and Oberman, are nothing less than tradition-breaking, with the methods of modern science being established to fill the void. Interestingly, however, it is at this point that agreement between the two breaks down.

For Leff, the important methodological developments that follow are those that bring mathematics and logical procedures into the study of nature.

> Now this detachment of physical theory from a predominantly metaphysical and theological context, in which it has been subordinated to wider metaphysical and theological issues, was the condition of its independent development; it enabled physical problems to be treated in their own terms by specifically physical and mathematical considerations and not as instances to illustrate metaphysical or theological questions. . . .[6]

The significance of this development for Leff is that

> . . . in the fourteenth century, for the first time, a distinctive body [of] scientific, mainly mechanical, theory arose with its own independent principles and procedures, which were self-contained and not subservient to higher nonphysical principles. It arose from the quantitative treatment, through the application of mathematics and up to a point logic, of problems which until then had been considered qualitatively under their different categories of movement, time, place, quantity, and quality.
>
> These new quantitative methods mark the beginning, albeit inchoate, of future scientific procedure, namely, in generalizing the conclusion arrived at by

calculation or analysis as formulae or laws to be taken as axioms on the Euclidean model and applied to physical phenomena.[7]

For Oberman, on the other hand, it is the experimental, not the abstract mathematical method, that develops in the wake of the *absoluta-ordinata* distinction.

> In both theology and physics the distinction between possibility and reality helped to free man from the smothering embrace of metaphysics. Yet in physics the same distinction works itself out in a different way . . . . Whereas in theology the established order . . . is at the same time the revealed order . . . , in the realm of physics the established order is the order of the established laws of nature, still to be investigated and freed from the babylonian captivity of metaphysical a priori.
>
> In this climate there emerges before our eyes the beginnings of the new science. We see the first contours of this science in a double thrust: (1) the conscious and intellectually ascetic reduction and concentration on *experientia* . . . [and] (2) the discovery of the scientific role of *imagination* that allows for *experiments*.[8]

Thus, some agreement seems to be emerging over the belief that the medieval period may be seen as paving the way for later developments, but with the questions of "for what?" and "how?" remaining very much open to debate.

That ambiguity emerges at this point is understandable. The division between Leff and Oberman reflects a deeper rift within the history of science itself over the cause(s) of the Scientific Revolution. If scholars could agree that Galileo or Newton was more an experimenter than a speculative scientist, or the reverse, it would then be easier to decide which of the many possibilities inherent in the *absoluta-ordinata* distinction might have paved the way for the Scientific Revolution. However, as long as debate continues over the best way to characterize and account for the Scientific Revolution, establishing a basis for judging the relevance of the Middle Ages to the history of science will remain a difficult and uncertain task.

The problem of relevance is, moreover, further complicated by the fact that it is not at all certain that any of the frameworks or models used to characterize the transition from medieval to modern has any explanatory power. Funkenstein operates within the traditional model adopted by intellectual historians. He views the history of science as the ongoing process of advancing increasingly sophisticated descriptions of nature, resting the importance of the *absoluta-ordinata* distinction on the contribution it makes to generating new ideas. Leff and Oberman operate more within the methodological model, adding the key element of a critical values transformation — from universalism to particularism — which creates the demand (or opportunity) for new methods.

The chain of events that develops under this model is first the shift in values (fourteenth century), which sets in motion the secondary chain of events (methological development, fourteenth–seventeenth centuries), which produces the Scientific Revolution.

As explanatory principles, both models have weaknesses. Neither can forecast the course of historical events. Knowing when an idea might first have been advanced, either as a tentative hypothesis or to be rejected, provides no information on when that idea will be adopted, *i.e.* cause a revolution. For the latter, Funkenstein readily admits one has to turn to other factors, such as "a different climate of opinion, tensions within the old theory, developments in other fields, [and/or] new factual evidence."[9] Leff and Oberman's model, even though it focuses on a more fundamental and potentially dynamic factor — a basic values transformation — does not account for the long delay between the transformation and its fruits. Once again, other factors would have to be brought in to explain why the potential that seems to be present in the medieval period took so long to develop.

The fact that it seems so necessary to turn to "other factors" to explain the timing and actual transformation from the potential of the medieval period to the Scientific Revolution strongly suggests, I would argue, that changes at the level of these "other factors" may *be* the Scientific Revolution. Certainly, it would seem worthwhile to test this hypothesis by looking in more detail at changes in these "other factors" over time, *e.g.* by tracing out the course of the shift from universalist (wholist) to particularist (specialized) thinking. The model that would be used for conducting such studies would assume that the essence of the Scientific Revolution is an underlying change in basic values, which is symptomatically manifest as, among other traits, changing methodologies, world views, occupational patterns, educational programs, support structures, interest priorities, and a host of other factors discussed in histories of science.

That there are complex and important values transformations taking place during the medieval period has long been recognized. Brian Stock's paper in this volume and longer study of literacy represent only one of many recent attempts to understand the priorities of persons who lived before the obvious great surge that produced the modern era.[10] Lynn White has on numerous occasions advanced hypotheses on how technological thinking and technological values slowly captivated the Western mind. In particular, White has attempted to turn the traditional top-down mentality of scholars on its head and to show how technological values slowly changed the way intellectuals at the top thought about learning.[11] If such studies were accompanied by a genuine recognition of their importance to a full understanding of the relevance of the Middle Ages to the history of science and technology, the extent of that relevance could, I believe, be delineated with much greater precision and meaning.

The importance of changing values during the Middle Ages did not escape the notice of those who were affected. As one commentator on the contemporary scene in the thirteenth century lamented:

> The muses are silent, confounded, repelled, as if numbed by the sight of Medusa. But why, you ask? If you are a real scholar you are thrust out in the cold. Unless you are a money-maker, I say, you will be considered a fool, a pauper. The lucrative arts, such as law and medicine, are now in vogue, and only those things are pursued which have a cash value.[12]

For those of such a practical mind, the way of the future was clear. So it was that one father advised his son who had written for more money to study the Bible: "This requires, as I was told, a great deal of money. Therefore, you would be better advised to audit the arts, . . . either physics or another lucrative science, because you will not gain great wealth if you pursue the ministry."[13] The reflection of values in these two comments and the potential impact on medicine, law, and the whole structure of learning could not be clearer.

So too have modern scholars recognized the importance of deep-seated transformations on the Scientific Revolution. Jacob Bronowski argues in his *Common Sense of Science*:

> We sometimes speak as if science has step by step squeezed other interests out of our culture, and is slowly strangling the traditional ways of thinking. Nothing of the kind. The Scientific Revolution in the seventeenth century was a universal revolution. Indeed it could not have been unless there had already been a deep change in the attitudes to everything natural and supernatural among thoughtful men. Puritanism in England and Protestant martyrdom on the Continent are the religious traces of that change; Marvel and Molière mark it in the arts, and Cromwell's revolution and the wars of Louis XIV are its political traces. Nor, of course, were these changes in the climate of mind without practical antecedents. At the bottom, all derive from the explosion of the rigid hierarchy of land and craft which was the medieval world. . . .[14]

For Bronowski, this "regress to first causes takes us too far from the Scientific Revolution itself." In my view, such a "regress to first causes" takes us to the very heart of the Scientific Revolution and deserves to become the focal point of considerably more scholarly attention. It is only by pursuing such studies that the true relevance of the Middle Ages to the history of science and technology can be established.

NOTES AND REFERENCES

1. Pierre Duhem, "Un précurseur français de Copernic, Nicole Oresme," *Revue générale des sciences pures et appliquées* 20 (1909):866–873; and *Études sur Léonard de Vince*, 3 vols. (Paris: A. Hermann, 1906–13); Charles Homer Haskins, *The Renaissance of the Twelfth Century* (Cambridge, MA: Harvard University Press, 1927); see also *Studies in the History of Mediaeval Science* (Cambridge, MA: Harvard University Press, 1924).

2. A.C. Crombie, *Augustine to Galileo: The History of Science A.D. 400–1650* (London: Falcon Press, 1952); and *Robert Grosseteste and the Origin of Experimental Science* (Oxford: Clarendon Press, 1953); J.H. Randall, Jr., *The School of Padua and the Emergence of Modern Science* (Padua: Editrice Antenore, 1961).

3. Quoted from Edward Grant, *A Source Book in Medieval Science* (Cambridge, MA: Harvard University Press, 1974), p. 48.

4. William J. Courtenay, "Nominalism and Late Medieval Religion," in *The Pursuit of Holiness*, Charles Trinkaus with Heiko Oberman, eds. (Leiden: Brill, 1974), pp. 26–66.

5. Amos Funkenstein, "On the Role of Hypothetical Reasoning in the Emergence of Copernican Astronomy and Galilean Mechanics," in *The Copernican Achievement*, Robert S. Westman, ed. (Berkeley, CA: University of California Press, 1975), p. 170.

6. Gordon Leff, *The Dissolution of the Medieval Outlook: An Essay on Intellectual and Spiritual Change in the Fourteenth Century* (New York: Harper & Row, 1976), p. 95.

7. *Ibid.*, p. 96.

8. Heiko A. Oberman, "Reformation and Revolution: Copernicus's Discovery in an Era of Change," in *The Cultural Context of Medieval Learning*, J. E. Murdoch and E.D. Sylla, eds. (Dordrecht: Reidel, 1975), pp. 408–409. The general thesis advanced by Leff and Oberman is also discussed by E.R. Woods, "Ockham on Nature and God," *Thomist* 37(1973): 69–87.

9. Funkenstein,[5] p. 170.

10. In addition to Stock's article in this volume, pp. 7–19, see *The Implications of Literacy: Written Language and Models of Interpretation in the Eleventh and Twelfth Centuries* (Princeton, NJ: Princeton University Press, 1983).

11. See "Medieval Engineering and the Sociology of Knowledge," *Pactific Historical Review* 44(1975):1–21; and "Medical Astrologers and Late Medieval Technology," *Viator* 6(1975):295–308.

12. Louis J. Paetow, ed., *Two Medieval Satires on the University of Paris* (Berkeley, CA: University of California Press, 1927), p. 155.

13. Charles Homer Haskins, *Studies in Mediaeval Culture* (Oxford: Clarendon Press, 1929), p. 25.

14. Jacob Bronowski, *The Common Sense of Science* (Cambridge, MA: Harvard University Press, 1953), p. 19.

The Mountain Men of the Casentino during the Late Middle Ages

JOHN MUENDEL

Department of History
University of Wisconsin–Waukesha
Waukesha, Wisconsin 53186

THERE IS A DEBATE among historians of technology regarding the proper orientation of their discipline. One school proposes that the history of technology should ultimately be social history—a discipline with an intimate understanding of technology, but fully aware of technology's relationship to society.[1] Another school of thought argues for "a broad and variegated discipline" which allows scholars from different fields, inside and out of academe, to contribute to its development.[2] Being very much involved in the social and economic history of the medieval period, I cannot help but favor the first of these convictions. However, my position must be tentative. Gustina Scaglia, for example, is an art historian who has made a significant contribution to the history of technology and, therefore, stands as an excellent example in support of the eclectic point of view.[3] This debate, then, may never be resolved, but I do think that there are certain methodological principles that historians of technology should follow whatever their opinions might be with respect to this question.

First and foremost, the work of an historian of technology should be empirical. Secondly, it should be regional and, finally, cumulative. Abbott Payson Usher has written much about the first and third of these categories. In short, he claims that explicit empiricism should be used on "a broader canvas" that allows for longer time intervals and greater flexibility of mind to integrate the many separate elements of technological development.[4] The second of these categories, however, needs greater emphasis. All too often generalizations are made about European medieval technology that are based upon assumptions that have not been truly tested in specific areas of the continent. For example, Fernand Braudel in his most recent work proclaims an "almost constant ratio" of European mills to population during the central and late Middle Ages to be 1 to 29.[5] Since the ratio of mills to population in medieval Pistoia was 1 to approximately 140 between 1310 and 1344 and the same again in 1427,[6] Braudel's figures, adopted from a study of Lazlo Makkai, are entirely misleading. Braudel also implies that during the central Middle Ages mills became a function of cities and their emerging exchange economy.[7] My recent studies of Florentine mills show that the development of the French mill, introduced into the territory of Florence in the middle of the thirteenth century, was purely a rural phenomenon with the effect of urban influence being at a minimum.[8] Medieval historians, influenced by sociological principles, might like to see the industry of the countryside suffering from the inefficiences of decentralization, but regional studies, particularly those of Italy, will prove them wrong and highlight the technological capacity of the medieval peasant.

As I have intimated, the region I have been studying is medieval Tuscany, Pistoia and Florence being two prominent areas of this region located in north-central Italy. In this presentation I would like to analyze, in particular, the social and technological development of still another area of Tuscany — the Casentino during the late Middle Ages.

The modern-day Casentino, located directly north of the city of Arezzo, is enclosed by mountain ridges on each of its sides: to the north and northeast by the Apennine divide dominated by the Monte Falterona and the Alpe di Serra; to the east by the Monti della Verna and the Alpe di Catenaia; and to the west and northwest by the Pratomagno and the Monte della Consuma. These ranges are interrupted only toward the south where the Arno, originating in the Monte Falterona, opens a significant passageway near the town of Subbiano, which can be considered the southernmost boundary of the area. During the medieval period, however, the Casentino was considered to begin at the point where the torrents Teggina and Archiano empty into the Arno west of Bibbiena. From here it extended across the Monte della Consuma to incorporate areas of the Valdisieve (see FIG. 1). During the Middle Ages, therefore, the mountain ridges of the northwest did not isolate the Casentino from the territory of Florence. In fact, only in the last century with the establishment of a railroad between Arezzo and Stia have the upper reaches of the Casentino been economically tied to the south.[9]

This area, dominated by castles found at heights approaching 600 meters, was controlled throughout the Middle Ages by the Guidi family. By the fourteenth century the Guidi from Battifolle were by far the most powerful segment of this clan. The contracts I have gathered regarding the leasing of hydraulic machinery in the area indicate that during the fourteenth century this branch maintained control from centers at Poppi, Pratovecchio, Castel Castagnaio and San Leolino, its ultramontane stronghold.[10] The power of the Guidi da Battifolle was dealt a serious blow on June 29, 1440 when the Florentine forces of Cosimo de' Medici defeated them and the Duke of Milan at Anghiari. Count Francesco del Conte Roberto II, the last representative of the Guidi da Battifolle, had to abandon his feudal domains and go into exile.[11] Florence,

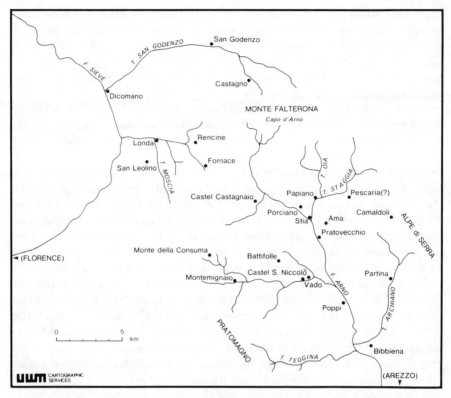

FIGURE 1. The Casentino during the Middle Ages.

however, did not take control of the whole area. As had been the case earlier in the fifteenth century, the feudal domains of the Casentino, save for Castel San Niccolò and parts of Pratovecchio, remained exempt from Florentine taxation through the last Catasto of 1480–81.[12]

Since the various Catasti shed considerable light upon the wealth and possessions of individual families living in the territory of Florence, their absence for a substantial part of the Casentino is a lamentable development for the historian attempting to recreate the social and economic structure of the area. Scholars, however, have resorted to other materials to achieve this end. Although she has mistakenly identified the grain mills of the area as possessing vertical waterwheels ("le pesanti ruote"), Rita Marignani has successfully employed the records of the abbey at Camaldoli to demonstrate that between 1000 and 1150 water mills acted as a secure index for the economic vitality

of the Casentino, particularly in the vicinity of Partina.[13] Philip Jones has analyzed the same source to show that even into the sixteenth century this monastery maintained firm control of its extensive possessions throughout the
Casentino by using a variety of contracts without ever relinquishing its de-
mesne farming of old. He indicates that large profits were also derived from
its livestock and dense forests of fir and beech.[14] Through the use of Florentine governmental records Richard Trexler has revealed that in the middle of
the fifteenth century the rivers of the Casentino abounded with trout which
the inhabitants poisoned with lime and nut shells so that quick catches could
be made.[15] Manuela Barducci has dealt with the fifteenth-century visitation
records of the bishop of Fiesole, which show that the strength of the Church
in the Casentino had been considerably weakened. Officials complained of
the poverty, misery, and ignorance of the area where the *contadini* felt no
need to construct or repair churches.[16] This may have been true. Notarial
records indicate that churches were controlled by lay communities who may
not have cared to support them.[17] However, as Jones' study suggests and as
I hope to show, the proclaimed poverty and misery were exaggerations.

These documents thus give some understanding of the way the society of
the Casentino functioned particularly during the fifteenth century. There is,
nonetheless, a source that is even more explicit for this period of time — the
account books of two ironworkers, Deo di Buono da Tracorte and his son
Giovanni da Stia. From them not only is a view of the daily activities of the
area obtained, but also considerable insight into the technological processes
which the region had inherited from its medieval past.

The journals, located in the Archivio di Stato of Florence, span the years
from 1458 to 1497. The debits of Deo's book cover the period between December 1, 1458 and August 29, 1465.[18] It was identified by the letter G and
acted as a continuation of his declarations found in a now lost book marked
with the letter F. The accounts of book G were subsequently transferred to
still another missing journal. This was a book of his son Francesco and identified with the letter A. In order to find that part of Deo's book G which did
not overlap with the others, I developed a cumulative index of all of his debit
entries. Assuming that book G was his principal, if not his only, work book,
this index revealed that the period from January 2, 1459 to January 29, 1460
contained approximately 84 percent of his entries and was, therefore, the period
of greatest validity for making statistical judgments regarding his output.

Since the accounts were subsequently transferred to the book of his son
Francesco, it must be assumed that Deo was in his declining years. Information from the surviving notarial chartularies of the period help to ascertain
specific details of his later life, especially with respect to his death. On Oc-

tober 10, 1463 Francesco and his brother Giovanni are found dividing the property of their deceased father.[19] The latest information regarding their father
prior to this date appears on April 3, 1461 when Deo, as one of the governors
of the hospital of San Iacopo and San Filippo at Pescaria, participated in the
selection of a new superintendent.[20] Deo, therefore, died sometime between
the spring of 1461 and the fall of 1463. Since the later entries of his book were
being transferred to Francesco's Book A, those following Deo's death were
presumably recorded by this son who remained an ironworker at Tracorte
until as late as October 28, 1475. On this date Francesco, "faber de Tracorte,"
sold the hospital of Santa Maria Nuova of Florence the very same grain mill,
saw mill, and site of a honing mill which he had received in the earlier division with his brother.[21]

This division of property indicates that Giovanni must have been the older
of the two brothers. A house with a shop located at Tracorte was divided between them with the stipulation that a wall inside the structure would be built
within 6 months obviously to make the division definitive. However, it is stated
earlier in the contract that Giovanni already possessed a house and shop located in the town of Stia and that all the furnishings which had belonged to
his father would become his own. Giovanni thus worked at Stia when he began
his surviving account book on December 1, 1466. His declarations continued
until September 23, 1476.[22] Since Giovanni's book, like that of Francesco, was
identified by the letter A, it must be assumed that he too was just beginning
his own independent accounts. Nonetheless, there are frequent transfers from
a non-existent book of debtors and creditors marked $\bar{a}$ which had also appeared
occasionally in his father's journal. It is not known how he used this book
beyond indicating accumulated debits and credits for his clients, but he could
have employed it to record specific transactions for work done in his shop.
Since he was also transferring later summaries to a lost journal marked B,
I, therefore, prepared another cumulative index to show that 87 percent of
his debit entries occurred between January 2, 1468 and April 22, 1472.

A second book of Giovanni's transactions is also extant. It was marked
with the letter G and, ironically, is contained in the second half of his father's
book identified by the same letter.[23] It begins on March 10, 1477 and continues to June 17, 1497. A cumulative index of the debits provides a period
of validity between January 2, 1478 and December 29, 1479 when 78 percent
of the entries occurred.

The style of recording utilized in the three books is what is usually referred
to as *alla veneziana*. In other words, Deo and Giovanni used the advanced
Venetian technique of accounting by registering the debit and credit entries
for each client on opposite pages of the same book.[24] Since they occasionally

referred to entries found in the account books of other individuals with whom they dealt,[25] this style was more than likely common in the Casentino and might indicate the progress of Italian capitalism in the more remote areas of the Apennine mountains during the late fifteenth century. Such a conclusion, however, would be entirely misleading. There can be no doubt that the information gathered from these books reveals a subsistence economy whose primary objective was survival, not profit. In the period between January 2, 1468 and April 22, 1472 only 25.5 percent of the credits were in cash payments of Florentine *larghi, soldi, quattrini,* and *denari.* All other payments were in work-days or *opere;* the produce of dispersed farmsteads and gardens; the products of local industries; and old, worn-out tools. The usual jargon of the urban merchant was certainly utilized by Deo and Giovanni, but it must be taken quite literally because none of their products were paid for directly. Thus, when *portò,* the third person singular of the past absolute of *portare,* was used, it did not mean "paid," but "carried away." Similarly, when *recò* was used, it did not mean "bought," but "fetched" as their sons retrieved goods from various parts of the countryside when the season allowed. While Deo and Giovanni were involved almost exclusively in the manufacture and repair of iron tools, they also made concessions of grain, meat, work, and cash to individuals in need, but in only one incident was a loan specifically implied.[26] If "interest" is to be inferred, it must be seen purely as a penalty for a payment long overdue. For example, in an undated entry occurring between April 5, 1471 and March 21, 1472 Giovanni reported that a certain Iacomo di Piero Marzocco from Ama "ought to give for a half bushel of grain granted to him throughout the year 1467, and up through 1471 he has not paid for two other bushels of grain." He then added in the margin beside this entry: "I have converted the said grain into money at a rate of thirty *soldi* per bushel."[27] Since the value of grain in the Casentino from 1466 through 1472 fluctuated erratically between 10 and 24 *soldi* per bushel, the specific amount of the penalty cannot be determined, but it was certainly significant even though Giovanni charged him for only two bushels.

The world of the Casentino during the fifteenth century was by no means entirely divorced from possible capitalistic endeavor. Giovanni himself, after completing the construction of a new shop at Porciano in December of 1470, built nearby a tavern that may have had behind it a profit motive.[28] The lumber cut by the saw mills of the area was tied together into rafts and floated down the Arno for an apparent profitable exchange.[29] The account books, moreover, indicate that fairs occurred particularly at Vado[30] while a market thrived, it appears, at Pratovecchio.[31] These could easily have acted as profitable outlets for the products of local industries. Nonetheless, taken as a whole, these ac-

FIGURE 2a. A relief of Andrea Pisano illustrating the shop of a Florentine blacksmith (from Giotto's Bell Tower, 1334–1337). Note that the smith is sitting, an unusual position for a man working iron. In the background are the bellows and the forge. To the right of the forge finished products (mattocks) are displayed. (Courtesy of Sansoni Editori)

FIGURE 2b. The stump and anvil in detail. The horn has been damaged. (Courtesy of Sansoni Editori)

FIGURE 2c. The smith's miscellaneous equipment in detail. (Courtesy of Sansoni Editori)

count books must not be seen as measures of urban influence in the mountains, but rather as sophisticated tally sticks which regulate an embedded economy depending almost entirely on reciprocation. Consider that when I calculated Giovanni's debit and credit entries for the period from January 2, 1468 through April 22, 1472, his debits equaled 1,199 *lire*, 1 *soldo*, and 9 *denari* while his credits amounted to 1,199 *lire*, 10 *soldi*, and 2 *denari*. That is a difference of only 8 *soldi* and 5 denari on the credit side. This stands as evidence for Giovanni's intuitive sense of balance in an economy essentially based on barter.[32]

The account books give no indication of the specific apparatus used in the shops of either Deo or Giovanni. However, leases for shops and equipment found in the notarial chartularies allow a composite picture of the typical forge to be drawn. This would include a tree stump for supporting an anvil, the anvil itself weighing some two hundred pounds and equipped with a steel working-plate, four legs, and a horn. Other items would include a pair of bellows for generating heat in the fire, at least four pairs of tongs, two hammers, a similar number of sledgehammers, a stone trough for quenching, a file, and a honing wheel for smoothing and sharpening finished products (FIG. 2). The accounts of Giovanni do show that he used honing wheels run by waterpower, but information about these devices will be given below.

It must be assumed that both Deo and Giovanni employed their sons to assist them in the manufacture and repair of tools. Francesco undoubtedly helped his father as Bartolomeo and Buono, presumably the oldest of Giovanni's sons, helped him. Nonetheless, Giovanni also made use of apprentices. In the fall of 1470 he did four days of work outside of his own shop with the aid of two of his *fanciulli*.[33] But if apprentices represented a source of labor from outside of the family, so did the employment of strikers or sledgers (FIG. 3). On May 17 and 18, 1459 one of the local millers, Paolo di Luca da Porciano, contributed two days of labor with the sledgehammer to work off part of his debt in tools and parts.[34] On four different occasions in February and August of 1467 Giovanni di Nencio di Meo da Stia did the same type of task,[35] an activity which any blacksmith will assert demands as much knowledge as his regarding the manufacture of iron.[36] The use of strikers from outside of the immediate family, therefore, indicates not only the common knowledge of the art, but also that sons could be relieved of certain duties within the shop so that other endeavors could be accomplished.

Perhaps the most striking feature of these account books is that Deo and Giovanni were supplied directly with new sources of iron and steel without the intervention of a middleman. Unlike the *fabbri* of the plains and middle hills of Tuscany who depended upon mercers or *ferraioli* to furnish them with

FIGURE 3. A Venetian smith being assisted by a striker or sledger (a relief from one of the arches of the Palazzo Ducale in Venice). (Courtesy of A. Kröner Verlag)

raw materials, Deo and Giovanni were within walking distance of the foundries. This provides a unique situation whereby Deo and Giovanni receive
their iron and steel in exchange for not only hatchets, hoes, and horseshoes,
but also for the manufacture of the specific parts of the founders' hydraulic
machinery. That there should be such a diversification in the manufacture
of iron should by no means be taken as a fifteenth-century or "Renaissance"
phenomenon. Since the late thirteenth century the Apennine foundries of Lucca,
Pistoia, Bologna,[37] and Florence had been receiving the iron ore of the Island
of Elba carried by barge and mule for its manufacture into transportable bars.[38]
The process of producing steel from wrought iron by carburization and
quenching was completely understood, moreover, by the Vikings of the ninth
century.[39] How the knowledge of this process came to Italy cannot be ascer-
tained. It may have needed no introduction since the Etruscan smiths of
Vetulonia knew the method as early as the seventh century b.c.[40] The Romans,
however, appear not to have fully comprehended the procedure.[41] Any connection with the Etruscans must, therefore, be considered doubtful. Be that
as it may, steel, as a separate commodity, was certainly used in the territory
of Florence during the Middle Ages as the history of the prominent Florentine
family known as the Acciaiuoli, or "steel-makers," makes evident. By 1341 it
had 53 factors spread throughout Europe to run its banking houses,[42] but the
family first settled in Florence during the twelfth century when it came from
Brescia to establish a foundry for the manufacture of carburized iron.[43]

The iron and steel manufactured in the Casentino was thus the primary
source for the production of tools by Deo and his son Giovanni. Between
January 13, 1469 and June 9, 1471 Giovanni himself received from Luca di
Cristofano delle Molina 446 pounds of steel while, during an equivalent period,
Biagio di Piero di Lorenzo da Stia supplied him with 1,118 pounds of bar iron.
Yet there were still other sources of iron and steel that must be taken into consideration before a full analysis of his raw materials during this approximately
29-month period can be achieved. First of all, Giovanni also received 165
pounds, 3 ounces of scrap iron from his customers. The economy of the Casentino, like that of Tuscany in general, was by no means one of waste. All tools,
no matter what their condition, had an inherent, regenerative value. In
Giovanni's shop even the metallic slivers that jumped from the red-hot iron
he was working had an essential worth. After oxidation they were swept up
and sold to Biagio for 20 *soldi* per hundred pounds in order to enhance his
production.[44] Secondly, Giovanni obtained 71 pounds, 4 ounces of iron whose
ultimate form was specifically designated prior to manufacture as well as an
undetermined amount of iron and steel advanced by individuals so that they
could receive a deduction for a finished product or a reparation. Examples

from these categories can be very informative. For instance, an iron wedge for splitting stone was made from two that were used;[45] a new chisel was produced from three that were worn-out.[46] It took, moreover, 9 pounds, 4 ounces of iron to manufacture a spade or *vanga*;[47] 20 pounds to make the heads of two axes;[48] and a 42-pound iron bar or *spiaggia* to fashion a saw for an hydraulic mill.[49] However, those entries that do not register a specific weight create a discrepancy in the overall amount of iron and steel utilized by Giovanni during this interval. Since most of the 36 entries comprising this category involved new or repaired agricultural equipment such as plowshares, colters, and mattocks, they cannot be summarily dismissed. The determination of any ratio between the amount of iron and steel necessary for the manufacture of tools must, therefore, be tentative, but a ratio of 3 to 1 seems plausible.

The independent production of iron and steel and this very index of raw materials amounting to three parts iron to one part steel point to an extremely important factor in the manufacture of medieval tools — the desirability of carbon-free, wrought iron. So often wrought iron is dismissed as a material that is too soft to be effectively utilized in a workable tool.[50] However, among the ferriferous metals it is the least subject to rusting; it can be easily welded; and, unlike steel, it will not burn during melting, thereby making it a reusable commodity.[51] The inherent economic value of wrought iron thus made "steeling," or the process whereby steel is welded onto an iron object, the predominant method of tool production during the Middle Ages.

Outside of the retrieval of scrap iron to which they attest, the account books of Deo and Giovanni give a firsthand presentation of this process. Let me once again use the entries of Giovanni's Book A from January 2, 1468 to April 22, 1472 as examples. During this period of time he made 651 new implements and repaired 990, a total of 1,641. The manufacture and repair of agricultural tools — plowshares, mattocks, spades, forks, billhooks, hoes, sickles, and scythes — represented approximately 41 percent of his output. These were succeeded by the production and reparation of the following items: shoes for asses, horses, and mules — 18 percent; woodsman's tools — axes, hatchets, pruning-hooks, knives, and mattocks for the preparation of charcoal piles — 14 percent; household fixtures — hinges, bolts, window guards, braces, spits, and tongs — 9 percent; carpenter's tools and equipment — adzes, chisels, planes, augers, gimlets, and nails — 6 percent; stone-mason's tools — mallets, picks, wedges, and chisels — 5 percent; parts for hydraulic machinery — 3 percent; tolls and equipment for the manufacture of woolen cloth — shears and components for spinning wheels — 2 percent; and, finally, miscellaneous articles — iron slivers, cooper's equipment, candle holders, and animal bells — 2 percent. Save for shoes, nails, spindles, household fixtures and miscellaneous articles,

FIGURE 4. A relief of Andrea Pisano depicting the use of the plow (from Giotto's Bell tower, 1334–1337). (Courtesy of Sansoni Editori)

all of these items were manufactured and repaired by the process of steeling. The principal part of the implement was wrought iron with steel being welded onto the cutting and piercing edges rather than those functional parts of a single piece of iron being worked to a desired hardness. The rest of the instrument remained comparatively soft. With use and age it could wrinkle, crack or peel, but it could always be regenerated by remelting and fusion with another worn-out piece, or by the process of patching. The steel parts also had to be treated, particularly for those instruments that needed a more durable working edge or surface. Anvils, wedges, stone chisels, pruning hooks, axes, hatchets, and hydraulic saws all required either softening through annealing or hardening by retempering. Nonetheless, by far the most common repair was patching and welding.

If the account books of Deo and Giovanni give a considerable amount of information regarding the specific methods of tool manufacture, they can go beyond such techniques into other fields of technology to shed further light on the social and economic character of not only the Casentino, but also of Tuscany in general during the medieval period. There are three such areas: the tillage of the soil, transportation, and types of hydraulic machinery. Space does not allow a full coverage of these subjects, but the salient points concerning each cannot be dismissed.

A general assumption among historians of Tuscan agriculture is that the heavy plow or *aratro* was not commonly used in most of Tuscany until well into the nineteenth century. In place of it men used the spade or *vanga*, which was extremely effective for the demands of dry farming but in the long run wasted time and energy.[52] The account books of Deo and Giovanni show, however, that two types of plowshare were prominent among their items of manufacture. One weighed from 9 to 15 pounds and was used on the *aratro* while the other weighed from 5.5 to 8 pounds and was employed on the light plow or *perticaro*. It must be taken for granted that in the mountains where the soil is dense and rocky a heavy plow would be more of a necessity than a spade. Yet the existence of two types of plows would argue for a greater dispersion of their use outside of the Casentino, especially since fourteenth- and fifteenth-century artistic representations of the plow show it to have been quite common (FIGS. 4 and 5). It is a subject that deserves more study, particularly with respect to the development of the *mezzadria* system which apparently contributed to improvements in agricultural techniques after 1350.[53]

A move from the debit to the credit side of these books brings us into the realm of transportation. One of the primary means of repaying a debt was haulage. Charcoal for burning in the forge, manure for spreading on fields and gardens, and wine for drinking in the tavern were carried by means of asses. Wood, stone, hay, straw, and grain, on the other hand, were dragged

FIGURE 5. A roundel of Luca della Robbia (1400–1482) illustrating the use of the plow (from the former study of Piero di Cosimo located in the Palazzo Medici). (Courtesy of the Victoria and Albert Museum)

FIGURE 6. A sledge or *treggia* at Caprese Michelangelo in the province of Arezzo. The photograph was taken by Paul Scheuermeier in 1924. (Courtesy of Longanesi & C.)

FIGURE 7. Two sledges at Cormezzano di Gassano in the province of Massa Carrara (1975). (Courtesy of Casa Editrice Bonechi)

exclusively on sledges (Figs. 6 and 7). A two-wheeled barrow used in the reparation of the church of Santa Maria delle Grazie[54] and another employed at the saw mill of Piero della Danza[55] were the only wheeled conveyances recorded in the account books of Deo and Giovanni. It must be assumed that there were few wheeled vehicles in the Casentino; the terrain permitted only those of small compass where their utility would be most effective. This condition lasted into the nineteenth century. When Emanuelle Repetti described the various areas of the Casentino in his geographical dictionary of Tuscany, a modern road was referred to as a "strada rotabile."[56] But again, such a circumstance was not an exclusive feature of the mountains. In 1369 when materials for the reconstruction of the dam of San Niccolò were brought to the city of Florence during a 4-month period, sledges were the primary means of transportation. Wheeled vehicles were not used at all.[57]

Leases found among the notarial chartularies tell much about the internal mechanisms of different types of mills. Since the individual hired to work a mill was responsible for all of its parts, they would frequently be listed and estimated either in the initial contract or the one terminating the lease.[58] For the period of the fourteenth century a number of such leases have survived for the grain mills of the Casentino owned by the Guidi da Battifolle.[59] During the fifteenth century contracts of this sort are nonexistent for any type of mill even though there remains a substantial number of chartularies depicting the activity of the Casentino. The explanation for this development is really quite simple: the owners worked their own machinery. In a situation analogous to developments in eastern Normandy between the fourteenth and sixteenth century,[60] the peasants and artisans of the Casentino had improved their standard of living at the expense of both the feudal and ecclesiastical authorities. As such, the leaseholder had become a self-employed proprietor. However, the void produced by the lack of notarial information is filled by the entries found in the books of Deo and Giovanni who manufactured and repaired parts not only for grain mills, but also for iron mills, fulling mills, saw mills, and honing mills.

A large amount of information is provided for the iron mill. Throughout the year 1459 Deo renovated the trip-hammers and bellows of Biagio di Piero di Lorenzo. His entries indicate the existence of pivots for the axle of his trip-hammers plus the iron bands that held the axle together.[61] He repaired the gate for the chute that fed the waterwheel of the hammers as well as the braces for the chute itself[62] and the counterweights for the bellows.[63] In 1468 Giovanni mended the tuyere or nozzle of Biagio's bellows[64] and, during those four days in the spring of 1470 when he did work outside of his shop with two apprentices, it was done patching the head of one of Biagio's trip-hammers and its

FIGURE 8a. The trip-hammers of the Papini mill at Maresca in the mountains of Pistoia (1978). (Photograph by author)

FIGURE 8b. A detailed view of the hammerheads and their anvils. (Photograph by author)

FIGURE 8c. A lateral view of the hammerheads and their anvils. (Photograph by author)

FIGURE 9a. The waterwheel of the Papini mill at Maresca. (Photograph by author)

anvil (FIGS. 8–10).[65] The water-driven trip-hammer, which freed blooms of slag incursions and shaped them into bars, was thus a vital part of the fifteenth-century Tuscan *fabbrica* as it had been since the beginning of the fourteenth century, if not before.[66] There is good reason to believe that its use in Europe may go back as early as the seventh century,[67] but all too often the employment of the trip-hammer during the medieval period is denied despite evidence to the contrary.[68] An outstanding historian of technology, A. G. Drachmann, when reconstructing anew the Roman oil mill described by Cato the Elder, included in the preface of his work the statement: "Old errors die hard."[69] One hopes that the aforementioned denial will have an easy death.

The pivot on which turned the axle of the trip-hammers helps to make a

FIGURE 9b. A detailed view of the waterwheel and its pivot. Note that excess water splashing off the wheel is caught by a makeshift funnel which leads it to the pivot to act as a lubricant. (Photograph by author)

probable determination of the type of fulling mill or *gualchiera* used in the Casentino during the late Middle Ages. Two sorts were known. One possessed vertical stamps moved by lugs or cams projecting from a horizontal axle while the other had an axle of the same inclination, but was moved by recumbent strikers. The *gualchiera* with recumbent strikers was the more efficient since its fulcrum absorbed the weight of the hammer rather than the rotating parts.[70] As I have stated elsewhere,[71] it appears that this type was predominant in the territory of medieval Florence. It seems that it was also the kind used in the Casentino. On two occasions Giovanni repaired the lugs of the fulling mill of Luca di Cristofano delle Molina.[72] However, on December 28, 1477

FIGURE 10. The axle of the trip-hammers of the Papini mill at Maresca. Note the lug or cam which strikes the end of the hammer to make it work. (Photograph by author)

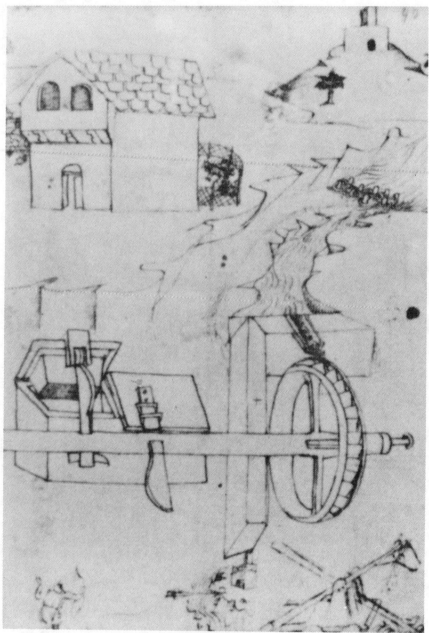

FIGURE 11. A probable fulling mill or *gualchiera* depicted by the Sienese engineer Mariano Taccola between 1419 and 1433. (Courtesy of the Bayerische Staatsbibliothek, München)

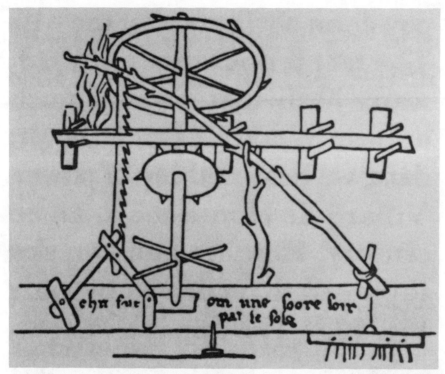

FIGURE 12. The saw mill of Villard de Honnecourt. The caption reads in translation: "How to make a saw operate itself." (Reprinted by permission of Harvard University Press, publisher of Abbott Payson Usher's *A History of Mechanical Inventions*)

he mended one of the pivots of this mill's axle by patching it with 6 pounds of his own iron and one of Luca's.[73] The two pivots or *aguigli* of Biagio's trip-hammers plus the three bands for securing the axle weighed 166 pounds.[74] A similar band bartered by Biagio weighed 17 pounds.[75] This would mean that the two pivots probably weighed more than 50 pounds each. Since the garnished pivot of the *gualchiera* was also an *aguiglio*, it was more than likely equivalent to those of the trip-hammers and functioned in a similar manner. This argument is by no means foolproof, but it does at least present the possibility (FIG. 11).

The saw mills of the Casentino were run by means of a crank and a connecting rod rather than the toggle joint featured in the thirteenth-century representation of Villard de Honnecourt (FIG. 12). Giovanni alone dealt with the equipment of at least five saw mills in the immediate vicinity of Porciano. In most cases saws were rejoined, patched, or sharpened,[76] but new saws were

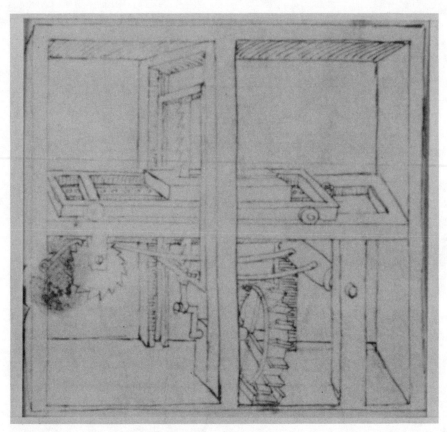

FIGURE 13. The saw mill of Francesco di Giorgio Martini (1489). Note that the means for attaching the connecting rod to the frame of the saw is obscure. (Courtesy of Edizioni Il Polifilo)

made, costing between 8 and 15 *lire*, by far the most expensive item in the account books.[77] The reparation of the *manfaro* or crank occurred twice, once for the mill of Bono di Bartolo Bono[78] and again for the mill of Duccino di Paolo da Papiano.[79] It is a significant find which gives greater credibility to the drawings of saw mills found in the notebooks of such Tuscan engineers as Francesco di Giorgio Martini (FIG. 13).

The use of the crank and connecting rod in the saw mill leads to still another problem — the mechanism of the honing mill. During the Carolingian period the famous long swords employed by the Frankish nobility were sharpened by stones turned manually by a crank as illustrated in the Utrecht Psalter

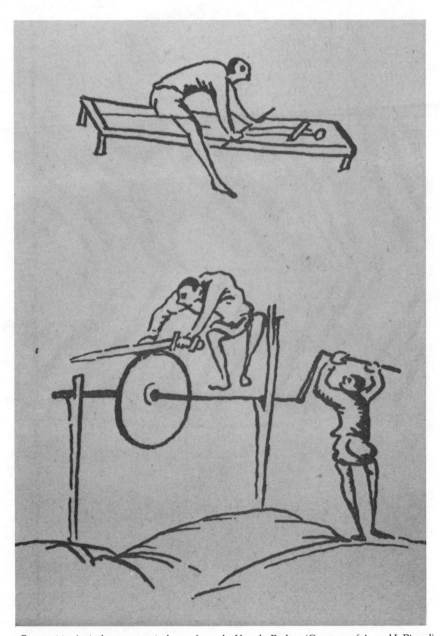

FIGURE 14. A ninth-century grindstone from the Utrecht Psalter. (Courtesy of A. and J. Picard)

FIGURE 15a. A reconstructed blacksmith's shop at the Palazzo dei Vicari in Scarperia (1981). (Photograph by Jeffrey Muendel)

FIGURE 15b. The interior of the blacksmith's shop at the Palazzo dei Vicari. (Photograph by Jeffrey Muendel)

FIGURE 16a. The grindstone or *rota* of the blacksmith's shop at the Palazzo dei Vicari in Scarperia. Note that the stone itself is of modest proportions. (Photograph by Jeffrey Muendel)

FIGURE 16b. The crank of the grindstone at the Palazzo dei Vicari. A billhook hangs on the wall in the background. The horns in the foreground are usually employed to carry whetstones for sharpening scythes. Possibly the smith used them to hold files. (Photograph by Jeffrey Muendel)

(FIG. 14). By the central Middle Ages the vertical *rota*, or wheel, for honing tools was commonly found in forges, particularly in Tuscany (FIGS. 15 and 16).[80] However, by the fourteenth century, if not before, this vertical grindstone was moved not only by hand, but also by water power. In 1358 a honing mill in the Tuscan province of Pistoia was run, oddly enough, by a horizontal waterwheel or *ritrecine*.[81] If the stone itself must remain in a vertical position to effectively hone instruments, the use of the horizontal waterwheel "throws a monkey wrench" into this development unless there was a means of converting the circular motion of the horizontal wheel into a similar motion on the vertical plane. My investigations in the Florentine archives during 1981–82 indicated that there was such a means. By the latter half of the fourteenth century Florentine artisans of the country side had successfully united the mechanisms of the horizontal and Vitruvian mills into one system to create the French horizontal mill, the "French" designating the interlocking gears of the classical Vitruvian type.[82] Thus when Giovanni "put to water" his honing mill for the forge at Porciano,[83] he very likely proceeded in the following manner: he employed a large-toothed wheel mounted horizontally on the axle of the *ritrecine* which transferred the proper circular motion directly to the honing stone by means of a horizontally set pinion, or lantern wheel. If the crank and connecting rod was a device well-known among the medieval sawyers of the Casentino, he could also have incorporated this mechanism in a way similar to a linkage devised by the Tuscan engineer Juanelo Turriano in the late sixteenth century for his notebooks dedicated to the Hapsburg king, Philip II (FIG. 17). This latter case is more speculative than the former, but it would not be going far astray to state that Giovanni, in either case, was continuing a tradition of technological skill derived primarily from the insight of the medieval, rural peasant.

The account books of Deo di Buono and his son Giovanni, therefore, make it possible to obtain a unique glimpse into the internal workings of a European mountain community during the fifteenth century. The view contains some apparent contradictions: a self-contained, barter economy linked to the business world by a cycle of fairs and markets; rural peasants owning and running expensive, hydraulic machinery without the intervention of the urban entrepreneur or the propertied aristocrat; and an advanced technology that does not use the wheel as a primary means of transporting goods. It is an economy that is, nonetheless, thriving because of the expansion of an iron industry which at the turn of the fifteenth century was in a period of recession.[84] The account books thus reveal a pre-industrial society working under optimum conditions.

Because of the overwhelming social pressure which propels or impedes it,

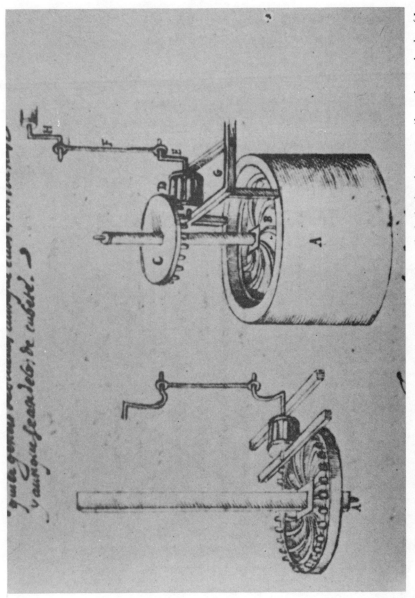

FIGURE 17. The crank and connecting rod activated by a horizontal, large-toothed wheel and its pinion (from the notebooks of Juanelo Turriano completed between 1554 and 1585). (Courtesy of the Biblioteca Nacional, Madrid)

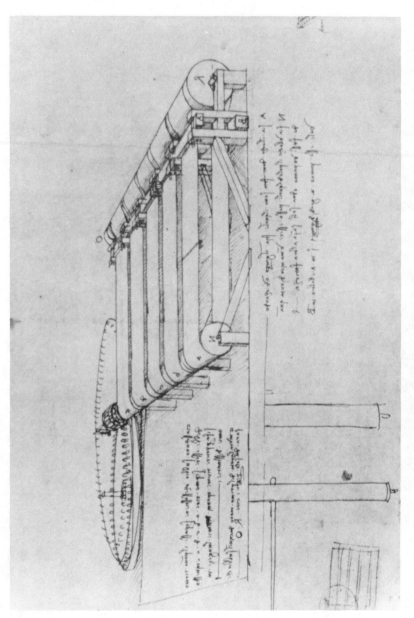

FIGURE 18. Leonardo da Vinci's gig mill incorporating the use of the large-toothed wheel and pinion. The gig mill was employed to raise the nap on shrunken pieces of cloth. (Courtesy of the Biblioteca Ambrosiana, Milan)

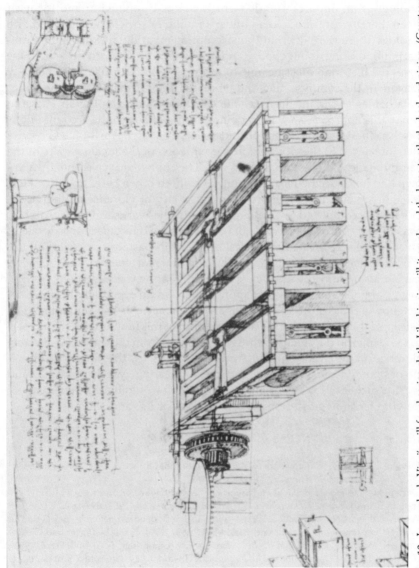

Figure 19. Leonardo da Vinci's mill for sheering cloth. Like his gig mill it employed the large-toothed wheel and pinion. (Courtesy of the Biblioteca Ambrosiana, Milan)

Fernand Braudel has recently expressed some doubt as to the validity of technology's role in history.[85] I hope that my study of these books has shown that technology is an important entity which can be examined in and for itself and can thereby contribute to an understanding of society. Within medieval technology there were internal transformations that did not depend on outside stimuli to bring about change and innovation. In order to discover these changes the historian must be able to examine, without prejudice, the daily endeavors of the common man and thereafter relate them to the more grandiose schemes of the intellect. Let me conclude by developing an image to facilitate this approach and, in so doing, bring us back to the account books themselves. Between the twentieth and twenty-fourth of April 1479 Giovanni di Deo journeyed to the city of Florence with one of the local sawyers of the Casentino, Duccino di Paolo da Papiano. The trip would have remained unknown had not Duccino lent Giovanni 23 *soldi* to buy a beret in one of Florence's shops.[86] At this time there lived in the city of Florence an individual who had just completed an apprenticeship in the workshop of Verrocchio. He was 27 years old and endowed not only with an extraordinary artistic talent, but also an intense curiosity. In April of 1479 this bastard son of a rural notary from Vinci was probably doing the experimental studies that would culminate in his painting of the Benois Madonna, or he was contriving his first drawings of guns and ballistic weapons which would eventually lead to a set of notebooks that future generations would perceive as the embodiment of the ultimate technological mind at work.[87] But at this juncture we should perhaps see Leonardo da Vinci and the two craftsmen from the Casentino meeting in Florence on the twenty-first of April 1479 to discuss the intricacies of machine design and the effectiveness of certain gearing systems (FIGS. 18 and 19). With such a vision historians of technology would have a better way of perceiving technological development and all of its ramifications.

NOTES AND REFERENCES

1.   David A. Hounshell, "Commentary on the Discipline of the History of American Technology," *The Journal of American History* 67 (1981): 854–865.
2.   Darwin H. Stapleton and David A. Hounshell, "The Discipline of the History of American Technology: An Exchange," *The Journal of American History* 68 (1982):897–900.
3.   Among other works see Frank D. Prager and Gustina Scaglia, *Brunelleschi: Studies of his Technology and Inventions* (Cambridge, MA: MIT Press, 1970); Frank D. Prager and Gustina Scaglia, eds., *Mariano Taccola and his book De Ingeneis* (Cambridge, MA: MIT Press, 1972); and Mariano Taccola, *De machinis: The Engineering Treatise of 1449*, Gustina Scaglia, ed. (2 vols.; Wiesbaden: Dr. Ludwig Reichert, 1971).
4.   Abbott Payson Usher, *A History of Mechanical Inventions* (2nd edit. rev.; Cambridge, MA: Harvard University Press, 1954), pp. 17–83.

5. Fernand Braudel, *Civilization and Capitalism, 15th–18th Century*, vol. 1: *The Structures of Everyday Life. The Limits of the Possible*, trans. Siân Reynolds (New York: Harper & Row, 1981), pp. 357–358.

6. See my "The Grain Mills of Pistoia in 1350," *Bullettino storico pistoiese*, 3rd series, VII (1972): 39–64 and "The Horizontal Mills of Medieval Pistoia," *Technology and Culture* 15 (1974): 194.

7. Braudel,[5] vol. 1, pp. 356–357.

8. See my "The 'French' Mill in Medieval Tuscany," *Journal of Medieval History* 10 (1984). In press.

9. Pier Luigi Lavoratti, *Il Casentino: Studio di geografia regionale* (Rome: Nuova Tecnica Grafica, 1961), pp. 5–7 and 11–16.

10. Archivio di Stato di Firenze (henceforth, ASF), Notarile antecosimiano (henceforth, Not.), A92–A94, P405, and G385, 1349–1392, *passim*.

11. Ferdinand Schevill, *History of Florence from the Founding of the City through the Renaissance* (1936; reprint, New York: Frederick Ungar, 1976), pp. 359–360.

12. David Herlihy and Christiane Klapisch, *Les Toscans et leur familles: Une étude du catasto florentin de 1427* (Paris: Presses de la Fondation Nationale de Sciences Politiques, 1978), pp. 139–140. ASF, Catasto: Quartiere di S. Giovanni (1459–1481).

13. "I mulini ad acqua della zona casentinese fino alla metà del secolo XII," *Argomenti storici* VI-VII (1981): 22–50.

14. "A Tuscan Monastic Lordship in the Later Middle Ages: Camaldoli," *Journal of Ecclesiastical History* 5 (1954); 168–183. It should be noted that Mr. Jones incorporates in his study the Catasto records of 1427–30 which do exist for this abbey.

15. "Measures against Water Pollution in Fifteenth-Century Florence," *Viator* 5 (1974): 455–467.

16. "Il Casentino nella prima metà del Quattrocento," *Argomenti storici* VI-VII (1981): 80–111. Cf. F. Coradini, ed., *La visita pastorale del 1424 compiuta nel Casentino dal vescovo Francesco da Montepulciano (1414–1433)* (Anghiari: Tip. Tiberina, 1941).

17. ASF, Not., A137, February 11, 1459 and A134, XI, no. 9, December 19, 1462.

18. ASF, Reale Arcispedale di Santa Maria Nuova (henceforth, RASMN) 475(a). The specific location of Tracorte cannot be determined. Deo's invocation (1r) reveals that it was in the "corte di Porciano del popolo di Sancto Lorenzo."

19. ASF, Not., A134, XII, nos. 1 and 2.

20. ASF, Not., A134, X, no. 4.

21. ASF, Not., A139, no. 1822.

22. ASF, RASMN 474.

23. ASF, RASMN 475(b).

24. Raymond de Roover, "The Development of Accounting Prior to Luca Pacioli According to the Accounts-Books of Medieval Merchants," in *Studies in the History of Accounting*, A.C. Littleton and B. S. Yamey, eds., (London: Sweet and Maxwell, 1956), pp. 138–139. *Cf.* J. N. Ball, *Merchants and Merchandise: The Expansion of Trade in Europe, 1500–1630* (New York: St. Martin's Press, 1977), pp. 25–28.

25. ASF, RASMN 475(a), 7r, 23r, 30r, 41r, 42v and 474, 80v, 94r, 104r, 110r.

26. ASF, RASMN 474, 37v.

27. ASF, RASMN 474, 125v.

28. For the construction of his shop between October 21, 1468 and December 17, 1470 see ASF, RASMN 474, 53r, 19r, 24r, 34r, 62r, 71r, 75r, 98r, 102r, and 91r. For the entries regarding the construction of the tavern in 1471 see ASF, RASMN 474, 87r, 113r, 79r, 87r, 94r, 124r, and 143r.

29. ASF, RASMN 475(b), 81v–82r.

30. ASF, RASMN 474, 105r. For other fairs without a specific location see ASF, RASMN 474, 13v, 27r and 475(b), (86r).

31. ASF, RASMN 475(b), 78v–79r.
32. Karl Polanyi, Conrad R. Arensberg, and Harry W. Pearson, eds., *Trade and Market in the Early Empires; Economics in History and Theory* (New York: The Free Press, 1957). *Cf.* William H. McNeill, *The Pursuit of Power: Technology, Armed Force and Society since A.D. 1000* (Chicago: University of Chicago Press, 1982), pp. 21–23.
33. ASF, RASMN 474, 80v.
34. ASF, RASMN 475(a), 12r.
35. ASF, RASMN 474, 6r.
36. Alexander G. Weygers, *The Modern Blacksmith* (New York: Van Nostrand Reinhold, 1974), pp. 24–25.
37. ASF, Not., R146, 84v, June 17, 1298 and S348, 33r, April 17, 1344.
38. David Herlihy, *Medieval and Renaissance Pistoia: The Social History of an Italian Town, 1200–1430* (New Haven: Yale University Press, 1967), pp. 41–43, 156, and 174–176. *Cf.* Muendel, *Technology and Culture* 15: 210–214 and Alessandro Gherardi, ed., *Le consulte della repubblica fiorentina dall' anno MCCLXXX al MCCXCVIII* (Florence, 1898), II, 673. For the sale of 2,000 pounds of iron "in vergellis et spiagiuolis" see ASF, Not., T279, September 29, 1386 (no pagination).
39. H.R. Schubert, *History of the British Iron and Steel Industry from ca. 450 B.C. to A.D. 1775* (London: Routledge and Kegan Paul, 1957), p. 76.
40. Carlo Panseri, *La tecnica di fabbricazione delle lame di acciaio presso gli antichi* (Milan: Associazione Italiana di Metallurgia, 1957), pp. 19–33. *Cf.* A. M. Snodgrass, *The Dark Age of Greece: An Archaeological Survey of the Eleventh to the Eighth Centuries B.C.* (Edinburgh: Edinburgh University Press, 1971), pp. 213–217.
41. Schubert,[39] pp. 54–57.
42. John Larner, *Culture and Society in Italy, 1290–1420* (New York: Charles Scribner's Sons, 1971), pp. 25–26.
43. The most recent treatment of the early history of the Acciaiuoli is found in Nicolas Cheetham, *Mediaeval Greece* (New Haven and London: Yale University Press, 1981), p. 166.
44. ASF, RASMN 474, 27v, 39v, 80v, and 107v.
45. ASF, RASMN 474, 68v.
46. ASF, RASMN 474, 52v.
47. ASF, RASMN 474, 6r.
48. ASF, RASMN 474, 81r.
49. ASF, RASMN 474, 56r. See note 38 for the use of *spiagge* in 1386.
50. Theodore E. Wertime and James D. Muhly, eds., *The Coming of the Age of Iron* (New Haven and London: Yale University Press, 1980), *passim.*
51. Alex W. Bealer, *The Art of Blacksmithing* (2nd edit.; New York: Funk and Wagnalls, 1976), pp. 40–41.
52. Carlo Pazzagli, *L'agricoltura toscana nella prima metà del '800* (Florence: Leo S. Olschki, 1973), pp. 165–178.
53. Herlihy and Klapisch,[12] pp. 260–263 and 267–283. The latest treatment of the problem is found in Judith Brown, *In the Shadow of Florence: Provincial Society in Renaissance Pescia* (New York: Oxford University Press, 1982), pp. 62–63, 78–80, and 94–99.
54. ASF, RASMN 474, 68v and 96v (continuing account).
55. ASF, RASMN 474, 28v.
56. *Dizionario geografico fisico storico della Toscana* (1841; reprint, Rome: Multigrafica Editrice, 1969), IV, 668.
57. ASF, Capitani di Parte, numeri rossi 114 (CRIA 6360).

58. Muendel,[6] pp. 215–217 and note 8, *Journal of Medieval History.*

59. See note 10.

60. Guy Bois, *Crise du feodalisme: Économie rurale et demographie en Normandie orientale du debut du 14ᵉ siècle au milieu du 16ᵉ siècle* (Paris: Presses de la Fondation Nationale des Sciences Politiques, 1976).

61. ASF, RASMN 475(a), 17v (March 30 and April 6).

62. ASF, RASMN 475(a), 17v (April 30; May 4 and 7).

63. ASF, RASMN 475(a), 29v.

64. ASF, RASMN 474, 39v.

65. See note 33.

66. R. Caggese, *Roberto d'Angiò e i suoi tempi* (Florence: R. Bemporad, 1921), I, 520–528 and Muendel,[6] *Technology and Culture* 15: 210–214. Rolf Sprandel, *Das Eisengewerbe im Mittelalter* (Stuttgart: Anton Hiersemann, 1968), p. 105, claims that a trip-hammer existed south of Livorno in 1185. His source reads: "Dicit quod hic testis expulit quendam carbonarium qui erat de Castello Vecchio facientem carbones in capite Boteofuoro, in Planale, et mallium ferri ei abstulit." See Natale Caturegli, ed., *Regesto della chiesa di Pisa*, vol. 24 of *Regesta Chartarum Italia* (Rome: Istituto Storico Italiano per il Medio Evo, 1938), p. 433. Even though information follows regarding a *fabbrica* in Rivo de Loreta, I must reserve judgment in identifying the above "mallium ferri" as a trip-hammer. For the sale of iron bars in 1386 see note 38.

67. Walter Horn, "Water Power and the Plan of St. Gall," *Journal of Medieval History* 1 (1975), 242–245.

68. Schubert,[39] pp. 126–138, but particularly pp. 137–138; Herlihy,[38] pp. 42 and 176; Norman J. G. Pounds, *An Economic History of Medieval Europe* (New York: Longman, 1974), pp. 320–326; and Ronald F. Tylecote, "Furnaces, Crucibles, and Slags," in *The Coming of the Age of Iron*, Wertime and Muhly, eds.,[50] pp. 221–222.

69. *Ancient Oil Mills and Presses* (Copenhagen: Levin and Munksgaard, 1932), p. 3.

70. Joseph Needham, *Science and Civilization in China*, vol. 4, part II: *Mechanical Engineering* (Cambridge: Cambridge University Press, 1965), p. 394.

71. See my "The Distribution of Mills in the Florentine Countryside during the Late Middle Ages," in *Pathways to Medieval Peasants*, J. A. Raftis, ed. (Toronto: Pontifical Institute of Medieval Studies, 1981), pp. 104 and 114, n. 108.

72. ASF, RASMN 474, 93v and 475(b), 52v.

73. ASF, RASMN 475(b), 26v.

74. See note 61.

75. ASF, RASMN 475(a), 17v.

76. ASF, RASMN 474, 28v, 34v, 84v, 113v and 475(b), 16v, 17v, 70v.

77. ASF, RASMN 474, 34v (13 *lire*, "Ischonto el suo ferro d'achordo chollui."), 131v (8 *lire*, ". . . fatta di suo ferro."), 134v (11 *lire*) and 475(b), 82v (15 *lire*).

78. ASF, RASMN 474, 61v.

79. ASF, RASMN 474, 115v.

80. Giulia Camerani Marri, ed., *Statuti delle arti dei corazzai, dei chiavioli, ferraioli e calderai e dei fabbri di Firenze (1321–1344)*, vol. VI of *Fonti sulle corporazioni mediovali* (Florence: Leo S. Olschki, 1957), pp. 128–129, 178–179, and 205–206. For the use of the crank see ASF, Not., I58, 75r, February 4, 1342 ("Item unam sofficem ferream et duo tagliaturia ferrea et unam ruotam lapidam cum uno manubreo ferreo.")

81. ASF, Not., N128, 85v, June 1, 1358.

82. Muendel.[8]

83. ASF, RASMN 475(b), 23r.
84. Muendel,[71] pp. 104–105.
85. Braudel,[5] vol. 1, pp. 430–435.
86. ASF, RASMN 475(b), 83r.
87. Kenneth Clark, *Leonardo da Vinci: An Account of his Development as an Artist* (2nd edit.; Harmondsworth, Eng.: Penguin Books, 1967), pp. 17–36 and 46–47.

Artisans, Drudges, and the Problem of Gender in Pre-Industrial France

DARYL M. HAFTER
Department of History
Eastern Michigan University
Ypsilanti, Michigan 48197

GENDER TABOOS WHICH define men's and women's work in different spheres are among the most basic elements in the work world. Rooted in the conservative practices of traditional folkways, separation of work according to gender was intrinsic to the corporate nature of pre-modern society. This distinctive feature of pre-industrial social structure was lost when modern industrialization downgraded individual craftsmanship and homogenized work and its professional formation. Yet even in the traditional era, the separation of task had exceptions. In certain cases early French society accepted without question homogeneity of work when it was justified by special status, privilege, or the lack of it. Far from showing its modernity, gender-free work situations confirmed the organic nature of their communities and the particular meaning that gender had for them. This study will explore the categories of craft work in pre-industrial France to show how the institutions of the Old Regime mediated and reorganized the traditional patterns of work.[1]

The largest group of undifferentiated workers in the middle ages were the impoverished. With regard to work, the desperately poor of France had lost their gender identity by virtue of their miserable condition. Men and women competed for unskilled jobs in every corner of the economy. They were alike in being deprived of access to technology and literacy. This debarring defined their position at the edges of society and kept them in it. It was as if poverty provided them with a separate status outside the boundaries of communal norms and expectations. Having no entrée into the rituals and standards by which technology was integrated into social life, the usual gender definitions that arranged work activities were irrelevant for them.[2]

Illiterate and migratory, the poor did not preserve the history of their occupations in a self-conscious record. Instead the survey of their employment has been reconstructed from the records of poor relief, police files, official reports, and church papers. Along with animals, they did all the back-breaking labor

FIGURE 1. Miniature of The Fables of Bidpay, fifteenth-century manuscript. Wood carriers in rags supplying the urban center. (Courtesy Editions de la Courtille)

FIGURE 2. The Virgin and St. Elizabeth spinning with a distaff and a swift, Master of Nuremberg, *ca.* 1400. (Courtesy Abner Schram Publishing Co.)

that the pre-industrial economy did not secure from wind and water power. Much of it involved transporting material from one place to another. Thus women worked alongside men carrying bundles of kindling wood (FIG. 1), sacks of grain, baskets of vegetables, and vessels of coal. They were among the construction laborers pictured in illustrated manuscripts, lugging stone, pitch, cement, and timber to the construction sites of cathedrals and secular buildings. They found employment in agriculture as wives of farmers or day workers. They worked as barge and boat women, ferrying clients from one bank of the river to another. In a more bizarre occupation, some waded out to meet ships offshore and carried passengers to the beach on their shoulders. They worked as unskilled servants in every kind of industrial enterprise, from foundaries to cloth-making firms. It was only when the job was concluded or in the absence of work that female qualities of these workers intruded into the transaction. In the first instance, their pay was approximately half that of their male coworkers. In the second, their resources in the "economy of makeshifts," as Olwen Huften has called the fragmented life of the poor, included occasional prostitution.[3]

These laborers at the bottom of the social scale can be contrasted with segments of the working poor whose livelihood was augmented by piece work.

Throughout the rural areas of France, thousands of women and men produced goods in a cottage industry system that was dominated by merchant entrepreneurs. The importance of their production can be seen in the estimate that a good proportion of industrial production came from rural householders.

Great variety characterized the conditions of their work. Some, like lace makers or knitters owned their own means of production; others were lent spinning wheels or looms by the entrepreneurs who deducted their cost from wages. Trades like shoe making and hat plaiting parceled out separate functions to the semi-skilled rural workers, while they brought the pieces into the city for finishing by skilled tradesmen. Everywhere the merchants' agents distributed raw materials to the workers who tried to augment their slim earnings by stealing portions and falsifying weights. A constant battle went on over the linen thread delivered for lace making, the silk cocoons assigned to be unwound and spun into thread (FIG. 2), the pieces of leather meant to become parts of shoes or saddle equipment. Moreover in the farm communities where the work was done, women did not have a monopoly on this "home work." Men could be found knitting, spinning, plaiting, weaving. But it would be a mistake to think that this involvement signified androgynous sharing of tasks. Technology was still highly sex-specific in domestic industry, but as Ivan Illich put it, "it was different in each valley."[4] The maxim that where one finds technology, one also finds gender control, still held.

This generalization becomes more problematic in regard to the free urban crafts outside the guild structure. Much town manufacture was done by unorganized workers, whose home was their workshop and whose families were their auxiliary workers. Women and men also worked on their own in a variety of trades unconnected to their spouse's craft. Despite such exceptions as the fourteenth-century well-paid mason's helper, most women in free trades performed designated work. Even when they worked side by side with men, they were usually doing that portion of the task set aside for women.

The absence of female names from the rolls of carpenters, roofers, glaziers, and other specialized construction trades illustrate the expected separation of task. Evidence dating from the thirteenth century shows that most women in the free crafts clustered in the production of textiles and the needlework trades. But we should be careful when assigning a gender-laden reason for this constellation of work; the predominant medieval industry was production of cloth and articles of apparel were among the most frequently purchased manufactures (FIG. 3 and 4). Furthermore the family involvement in some trades and the practice of widows carrying on their husbands' business, led to a blurring of clearcut activity. While sex-defined tasks were held as customary norms, urban industry tended to be less rigid in practice.[5]

FIGURE 3. Sculpture, north portal archivolt, Chartres Cathedral. Woman hackles flax to clean it. (Courtesy Abner Schram Publishing Co.)

FIGURE 4. Life of the Humiliaten orders, thirteenth-century manuscript. Woman reeling thread to form a warp. (Courtesy Bucher, Luzern/Frankfort-am-Main)

The mixing of men and women in the same crafts may be considered a harbinger of the capitalist impulse that encouraged the separation of gender from work. Like the presence of money in medieval towns, mixed work may have been both a symptom of a dynamic which encouraged society to move out of its medieval balance or an irritant helping to effect that transition. In the twelfth century, the feudal rural areas still depended for their artifacts on the manoral workshops and the *gynaceas*, both of which relegated work to specific sexes.[6] By the fourteenth century, the increasingly sophisticated taste of the nobility for cloth, adornments, farming tools, and armaments could only be satisfied by the more specilized technology of urban manufacture.[7]

An increasingly commercial market encouraged production outside the practices of customary professional formation. Encouraged by merchant entrepreneurs willing to advance capital for tools and raw materials, workers gravitated to the trades for which there was demand. Already-established artisans set up piece workers in rooms around the city, giving them quick instructions in techniques. While the flourishing crafts drew many workers of both sexes from the countryside, the pronouced migration of females to urban centers was notable.[8]

The presence of a large group of women employed alongside men in the free crafts brought unexpected pressures to bear in the workplace. Ross Davies' assertion that harmonious conditions between male and female workers in the middle ages came about because specific tasks were performed exclusively by one sex is not borne out by the facts.[9]

Men's occasional hostility to women's work hinged on the differences in rewards and behavior which were already visible in the most egalitarian social level, the mixed work of the very poor. Two factors exacerbated this problem: the lower wages women received and the different actions and conduct that set women apart and which inculcated a range of misogynist reactions. Reduced wages for women in mixed work originated spontaneously, but they came to be reinforced by official sanction.[10] Despite the difficulty in estimating comparative wages, the relation of female to male pay for the same work has been gauged at 3/5 or 3/4 in the fourteenth century and at 1/2 in the fifteenth century, while in the sixteenth century it fell even lower.[11]

Given that women, like men, were expected to earn enough to support themselves within the family, the wage differential frequently had a devastating effect on female workers. Already at a financial disadvantage, women workers were pushed off the edge of subsistence in the dead season or when business cycles caused orders to be cut. The occasional prostitution they resorted to in these instances concerned not only the public authorities; it also discredited them in terms of their normal craft activities. In a concrete way the lore of the craftsmen attributed the qualities of "the other" to them. At times they were accused of being frequently absent and indifferent workers. Metalworking activities were said to suffer from their magic capacities to "turn" the casting process. Where they were not accused of spoiling the product, they were sometimes blamed for spreading sickness to other workers.[12]

Complaints such as these are recorded in situations when women were involved in the introduction of new technologies. Far less likely to be workers in the regulated crafts, available at a lower wage, and more vulnerable than men, female workers were likely candidates for the work force of new industries or recently introduced processes. Their choice of employment had a significant effect on the development of rural industries. The most dramatic example of a new female-dominated industry is the eighteenth century introduction of cotton spinning in the area around Rouen, and its even more striking proliferation of English spinning machines at the end of the century. Characteristically, the woolen makers of Rouen tried to force the Intendant to prohibit these industries. He withstood their pressure, fearing that the women would descend on the city for help if they were deprived of work.[13]

The free trades operating in most towns and villages in France were curtailed by royal imposition of state industries and guilds in the fifteenth and sixteenth centuries. Making official the arrangements that had been informal practices, the legislation took account of women workers. It sought to preserve places for them in the new craft institutions. For instance, Louis XI decreed in 1466 that the new silkworks he was fostering in Lyon should be free to all workers. The law read: "Each [craft] may be undertaken by men and women of every classification, including the ten thousand persons from the city as well as the environs, and including [individuals] in the Church, nobles, female members of orders as well as others. . . ." Certain auxiliary functions like unwinding the silk cocoons and twisting the thread on a doubling frame were reserved for women. The woolen cloth industry in Saint-Omer and other cities also reserved functions accessory to the male weavers for female workers.[14]

While royal power could establish new industrial concerns and reserve traditional occupations for women in them, the king's attempt to neutralize the element of gender in work was ineffective. That monumental feat was done by admitting women into guilds on their own right. Transforming free workers into guildwomen conferred upon them the license to be treated solely as economic beings. For women's guilds, the advantages of reaping the profits of master, providing professional training, and undertaking businesses on the same terms as men must have been highly desirable. Acting as guild officers where that was legal gave these women experience in negotiating with the municipal and central government that was rare in the Old Regime. Mixed guilds were founded on the professional equality of the sexes; many women advanced to the privileges of mastership. In some, women took their places as jurées next to the men. In both, the guild's legal privileges and its coherent structure ensured that the training, products, and wages of female apprentices, journeymen, and masters would in principle be the same as men's.[15]

The guilds and other institutions which had the capacity to remove women's economic and legal disabilities benefited from the Old Regime's practice of allowing private groups to abrogate special powers and privileges to themselves. These included the privilege of being taxed as a group, the authority to make formal protests about royal policy affecting the corporation, and the exclusive right to perform special functions — legal in the case of parlementarians and technical in the case of guild members.[16]

Most early guild records in France date from the thirteenth and fourteenth centuries when free crafts in urban areas petitioned the king for permits enabling them to become incorporated. Incorporation gave them control over their technology as they imposed training and examination of masterpieces on apprentices. In this testing ground, new instruments and processes were examined. The successful ones were validated and the stamp of approval given for the award of municipal or state subsidies. Guild regulations, later reworked as laws for the kingdom, gained legitimacy and guild officials could appeal to municipal administrators for their enforcement. A technological monopoly that ensured high standards in the execution of industrial processes could be maintained in urban areas. The unpaid apprentices' work and their fees provided capital for the masters' industrial investment; that capital would become increasingly important in terms of export trade and the distribution of luxury goods in which France excelled.[17]

The guilds also institutionalized medieval conceptions of protection and sanctity for workers, apportioning resources and enforcing the just price. Dedicated to its own patron saint, each organization observed a festival day with processions that included banners, special masses, and prayers for members' well-being. The groups functioned as burial societies and collected funds for widows and children of their members. Their officers visited the workshops to maintain a seasonal schedule of hours and to enforce adherence to production rules and prices. Through the agency of the guild, the hierarchy of apprentice, journeyman, and master was formalized, although new evidence shows that unaffiliated valets and servants also worked in the *ateliers*. The guild was the legal entity by which workers exerted influence on their town governments. Because of their incorporation, they were able to mediate the fiscal demands of the royal authorities as well, protesting the sum of their taxes and apportioning them to their own members.[18]

A debate has arisen over the nature of women's involvement with guilds.

The overriding questions asked by social historians have been whether women were admitted into the *cursus honorum* within the guilds or whether they were subsumed under the status and activities of their husbands; were they bona-fide apprentices or merely auxiliary workers whose status was ensured by family ties to members of guilds, or by separate adherence to crafts set aside for them?[19] Linked to these queries has been uncertainty over the extent to which women governed themselves with their own officials, and the separate issue of the status of widows and daughters. The problem is compounded by its size and by the variety of rules and adjustments that were made in the process of craft incorporation. Economic pressures through the course of years also made for changes in rules and practice. Many customs were unwritten or were expressed unclearly in the statutes. While patterns of behavior may be found throughout certain crafts or among a group of cities, the element of variety remains to bewilder researchers. Documentation from the restrictive policies of certain English guilds casts further uncertainty over the likelihood of French women's full participation in the corporate organizations. The paternalistic and hierarchical nature of guilds suggests that they were not fertile areas of female participation.[20]

In the face of conflicting ideas, the inquiry turns to the documentation that has been assembled. Guilds that arose spontaneously in the thirteenth and fourteenth centuries had two designations for women: crafts that were mixed and those that were exclusively female. Records show a wide range of skills for both types of corporation. Indeed the surprising circumstance for the scholar of modern economic activity may be the extent to which men joined the women in making delicate luxury articles. In Paris the mixed guilds included embroiderers, makers of hats with gold and pearl embroidery, spinners of silk, felt makers and makers of rosary beads, makers of hats with peacock feathers, linen workers, furriers, and merchants of cloth at the central market.[21]

The Paris tax roles for the *taille* of 1292 and 1300 list a set of workers in crafts undertaken exclusively by women: unwinding thread from silk cocoons and making it into skeins, combing the silk, combing wool, carding wool with large spikes or wooden plaques studded with iron nails, joining the strands of wool, winding the warp onto looms, hairdressing, making silk thread bobbins for shuttles, and sewing *aumônières sarrazines*, brightly colored purses originally intended for the use of crusaders collecting alms on their way to the Holy Land.[22]

Much more study will need to be done before we have a comprehensive and accurate picture of guildwomen in Europe, but we know that when free trades were incorporated, they generally described their members as *maîtres et maîtresses, ouvriers et ouvrières, apprentis et apprenties*. Henri Hauser wrote that "the names of these workers of both sexes are even very often written out in the founding ordinance."[23]

The fact of women's presence in the guild hierarchy is also shown by laws designed to exclude them from certain work processes and from recognized mastership, apprentice, and journeymen's rights. By statutes in 1581, 1597, and 1602, royal authority generalized guild law, pressuring many free trades to become incorporated. While some large towns were able to resist the imposition of the *maîtrise* temporarily, powerful trends in the sixteenth century imposed the guilds throughout the kingdom as fiscal and economic aids to royal power. In the course of the reorganization, the amount of auxiliary help a master could keep was restricted and the status of women was specifically controlled.[24]

Several factors combined to diminsh women's access to guilds in the fifteenth and sixteenth centuries. Royal efforts to extend control over the kingdom more thoroughly led to the explicit contraction of women's legal rights. Women were now prevailed upon to make contracts only with their husbands' permission while widows were restricted in their use of the property they had inherited.[25]

Natalie Zemon Davis has reasoned that economic troubles may have been behind the sixteenth-century restrictions imposed by silk merchants and masters, limiting the number of male apprentices at weaving to two, outlawing female apprentices outright, and restricting the work of journeywomen. In the increasingly capitalist formation of Lyon's business community, women were being deprived of the possibility of mastership and ushered into the role of less skilled workers.[26]

A new emphasis on property inheritance by male heirs was another pressure dealing women out of the guild hierarchy. In the course of the sixteenth century, women lost their right to participate in the prestigious gold and silversmith trades. Lesser crafts as well found reasons to exclude women who had been *maîtresses*. Davis reports that the Lyon barber-surgeons guildmasters suppressed the letters of mastership owned by women in the trade in the mid-sixteenth century in return for a small monetary compensation.[27]

Complaints against women stood out in the disenfranchisement process. The Lyon barber-surgeons faulted the women for renting rooms to journeymen barbers whose incompetence caused "accidents and misfortunes." Women involved in food sales, the merchandising of old clothes, or lending money were held in general disrepute. They were finally accused of placing sneak thieves in homes and of fencing stolen goods. Paris linen makers were among the women accused of prostituting themselves.[28]

Exclusion from the guilds meant that the skilled female workers were prohibited from practicing the more lucrative trades. To justify this separation, supporters of the restrictions adduced special physical disabilities that made certain work harmful to women. Weaving, an activity that had been carried on regularly by men and women, was forbidden to females in the silk trade in the fifteenth century. The introduction of the draw loom, they claimed, brought difficult and complicated processes suited only to men. Instead, women were admitted into the workshops to make up the bobbins, read the complex designs from color coded charts, make up the bundles of threads needed for the draw looms, all of which were physically easy tasks; however, they also used their hands to dip the silk cocoons into boiling water to begin the process of unwinding; and as drawgirls, they tugged down cords which were fastened to large stones keeping the looms stable, an activity which customarily broke their health by the age of twenty-one. Fulling was also a trade that came to be relegated solely to men because women were judged too weak, despite their earlier involvement in the labor.[29]

These examples tend to support the argument that when society separated women from the exercise of particular technologies, it brought those crafts under the aegis of male jurisdiction and custom. Or to put it another way, crafts long pursued by men and women alike were labeled as gender-related. The guilds' function as an institution that transformed workers into "economic" individuals became irrelevant to those workers who were expelled from the mastership. These displaced workers suffered loss of wages and status because they no longer enjoyed the protection of the guild against the competition of other workers or against the economic disabilities inherent in women's lot.

A few women benefited from the designation of certain crafts as natural to women. The Paris needle trades saw the introduction of seamstresses into the guild to make clothes for women. During the course of the seventeenth and eighteenth centuries, the women competed with their male colleagues and the authorities protected them. The *"cheve" d'atelier* Catherine Gallopine and her girls secured the right to clothe the royal children. As a consequence, they were formed into a guild in 1675. Their subsequent controversies with tailors were settled with the decision that seamstresses would make undergarments for women and tailors those for men. At Dijon, the seamstresses managed to beat their male rivals for a monopoly of sewing. At Saint-Omer, a magistrate ruled in 1612 that the *couturières* could make garments for children and repair and refresh old clothes.[30]

These examples show the good fortune of a few women whose corporate organizations grew stronger in the modern era, but they were not alone. Despite the decline in women's guilds, in their legal autonomy, and in their wages, a number of women continued to participate in guilds that were composed exclusively of women or were of mixed membership.[31] While we need many more case studies of female guild life in the eighteenth century, the information available offers a strong case for the persistence of some women's guilds until their suppression in 1791. These guilds were able to endure because of the particular circumstances of the communities in which they existed. The evidence allows one to speculate that they found a more encouraging climate in the commercial atmosphere of northern France, where independent feminine entrepreneurship lasted into the nineteenth century.[32] Among them can be noted the prestigious flowermakers guild composed of women, the linen makers, seamstresses, merchants of fashionable clothing, and plumemakers of Paris. In Rouen, guildwomen were spinners, hosiers, drygoods merchants and woolen cloth makers, fashionable plume makers, old and new clothes sellers, tailors, makers of gold braid, and seamstresses of religious vestments. Further research is needed to determine the extent of guildwomen in other cities, and the numbers of women who entered guilds after their reconstitution in 1776. An edict in that year opened all guilds to women who might become masters if they had passed through apprenticeship and the term of the journeyman, had reached eighteen years of age (it was twenty for men), and had paid the fees.[33]

The definitive suppression of guilds in 1791 struck at the essence of the Old Regime, by cutting off the sources of privileged funding that the king had used to circumvent constitutional requests for taxes. To a certain degree, the attack on guilds was inspired by the wish to open the most lucrative crafts to all. Women workers had always been the object of special concern in this regard. Throughout the rise of guilds, commentators had urged that some craft work be reserved for women to keep them from the exigency of prostitution. This theme received even more attention during the Enlightenment, when abstract ideas of natural rights called into question the technological monopoly of the guilds. But, like modern historians, these reformers ignored the achievement of a small group of women whose work and legal condition had been nurtured within the protection of the corporate system. Only by means of special sanction had the overwhelming disability of their sex been overcome. For them, the essence of privilege conferred legal and technological equality with men. A fragile shell, the protection of the small workers' elite fell away as privilege was stripped from noble and cleric as well. With an egalitarian regime in which every individual had the same relationship to the law, women artisans joined the crowd of unprotected workers, saddled with the additional disadvantages of lower pay and technological discrimination in a modernizing industrial setting.[34]

NOTES AND REFERENCES

1. See Diana Leonard Barker and Shiela Allen, eds. *Sexual Division and Society: Process and Change* (London: Tavestock Publications, 1976) for a discussion of legal changes relating to gender. Sherry Ortner, *Sexual Meanings: The Cultural Construction of Gender and Sexuality* (Cambridge: Cambridge University Press, 1981) gives an anthropological approach. See also Evelyne Sullerot, *Histoire et sociologie du travail féminine* (Paris: Gonthiers, 1968). Louise A. Tilly, "The Social Sciences and the Study of Woman: A Review Article," *Comparative Studies in Society and History* 20, no.1 (1978):163–173, comments on Michelle Zimbalist Rosaldo and Louise Lamphere, eds. *Women Culture and Society* (Palo Alto, CA: Stanford University Press, 1974) emphasizing that anthropologists have been slow to point out the male-oriented nature of society and its effect on women.

2. Natalie Zemon Davis, "City Women and Religious Change," in *Society and Culture in Early Modern France* (Stanford, CA: Stanford University Press, 1975), pp. 72–73 on female and male literacy in Lyon.

3. Fundmental to our knowledge of workers in history is Olwen H. Hufton, *The Poor of Eighteenth-Century France 1750–1789* (Oxford: Clarendon Press, 1974) and her "Women and the Family Economy in Eighteenth-Century France," *French Historical Studies* 9 (1975): 1–22. See also Davis,[2] pp. 70–71 and Joseph Gies and Frances Gies, "A City Working Woman Agnes li Patiniere of Douai; Women and Guilds," in *Women in the Middle Ages* (New York: Thomas Y. Crowell, 1978).

4. See Ivan Illich, *Gender* (New York: Pantheon Books, 1982) for a provocation discussion.

5. See Andrée Lehmann, *Le Role de la femme dans l'histoire de France au moyen age* (Paris: Berger-Levrault, 1952), pp. 435–436, 440–442; and Étienne Martin Saint-Léon, *Histoire des Corporations de métiers*, 3rd edit. (Paris: Félix Alcan, 1922), pp. 195–227.

6. Henri Pirenne, *Economic and Social History of Medieval Europe* (New York: Vintage, 1937), p. 81; and Lehmann,[5] pp. 195–199.

7. Edward J. Nell, "Economic Relationships in the Decline of Feudalism: An Examination of Economic Interdependence and Social Change," *History and Theory* 6, no.3 (1967):317 argues that consumer demand on the part of the rural nobility was the salient factor in transforming the feudal economy and promoting urban development, not the dynamic city acting on the countryside.

8. For aspects of women's migration see Abel Chatelain, "Migrations et domesticité féminine urbaine en France, xviii siècle-xx siècle," *Revue historique, économique, et sociale* 47(1969): 506–528. A. Poitrineau, "Aspects de l'emigration temporaire et saisonnière au Auvergne à la fin du xviiie siècle et au début du xix siècle," *Revue d'histoire moderne et contemporaine* 8(1961):23–24; and Dayton Phillips, *Beguines in Medieval Strasburg* (Ann Arbor, MI: Edwards Bros., 1941), p. 21.

9. *Women and Work* (London: Hutchinson, 1975), pp. 27–43. He claims that antagonism between male and female workers originated when easier machines drew women into workshops early in the modern Industrial Revolution.

10. Bronislaw Geremek, *Le Salariat dans l'artisanat parisien aux xiiie-xve siècles*, trans. Anna Posner and Christiane Klapisch-Zuber (Paris: Mouton, 1968), p. 91, reports that an ordonnance of Jean le Bon in 1351 fixed women's daily wage in the Paris area at 12 *deniers*. He also reports an exception to the law. Davis,[9] p. 29, cites the 1388 statute of Richard II which set the wages of women in England at a lower rate than those of men.

11. Henri Hauser, *Ouvriers du temps passée (xve-xvie siècles)* (Paris: Felix Alcan, 1909), pp. 158–159.

12. See Mircea Eliade, *The Forge and the Crucible*, trans. Stephan Corrin (New York: Harper

& Row, 1962), pp. 79–96, for information on traditional rituals in metallurgy. Accusations of women workers' infectiousness are discussed in Daryl M. Hafter, "The 'Programmed' Brocade Loom and the Decline of the Drawgirl," in *Dynamos and Virgins Revisited: Women and Technological Change in History*, Martha Moore Trescott, ed. (Metuchen, NJ: Scarecrow Press, 1979), pp. 60–61.

13. On the effect of women workers' influence on industrial evolution see Daryl M. Hafter, "Agents of Technological Change: Women in the Pre- and Post-Industrial Workplace," in *Women's Life Cycle and Public Policy*, Dorothy G. McGuigan, ed. (Ann Arbor, MI: University of Michigan Center for Continuing Education for Women, 1980), pp. 159–168. Gay L. Gullickson shows how the proliferation of mechanical spinning and the need created for additional weavers, drew men away from farming and opened this formerly male activity to women. The rural industry she describes relegated particular crafts to different sexes. She writes: "In the eighteenth century, men and women generally worked in different places at different tasks" (p. 193). See "The Sexual Division of Labor in Cottage Industry and Agriculture in the Pays de Caux: Auffay, 1750-1850," *French Historical Studies* 12, no. 2 (Fall 1981):177–199.

14. Hauser,[11] pp. 48, 149–150.

15. Gustave Fagniez, *La Femme et la société française dans la première moitié du viiie siècle* (Paris: J. Gamber, 1929), pp. 95–96; Hauser,[11] p. 152, pp. 143–146, cites the fifteenth century (1485) petition made by the prudesfemmes ("maitresses-jurées femmes et filles") of the linen workers' guild. This female guild was organized just as the men's guilds were, with some minor differences in having no limitation on apprentice numbers.

16. For an informed discussion of privilege and its meaning in the Old Regime see C.B.A. Behrens, *The Ancien Régime* (London: Thames and Hudson, 1967), pp. 46–62 and Pierre Goubert, *L'Ancien Régime* (Paris: Armand Colin, 1969), pp. 9–31.

17. For a skillful analysis see William H. Sewell, Jr., *Work and Revolution in France: The Language of Labor from the Old Regime to 1848* (Cambridge: Cambridge University Press, 1980), pp. 16–39.

18. Standard sources of information on guilds in France are Emile Levasseur, *Histoire des classes ouvrières et de l'industrie en France avant 1789*, 2 vols. (Paris: Arthur Rousseau, 1901) a revised edition of his *Histoire des classes ouvrières en France depuis la conquête de Jules César jusqu'à la Révolution*; Emile Coornaert, *Les Corporations en France avant 1789* (Paris: 1941); Saint-Léon, *Histoire des corporations*; Hauser.[11] It is now accepted that diverse tasks were performed by day workers called valets with no claim to guild status. Geremek,[10] pp. 51–66.

19. The familial structure of guild workshops was underscored by Geremek,[10] pp. 14, 20 *et passim*. See the fundamental analysis focusing on the eighteenth and nineteenth centuries by Louise A. Tilly and Joan W. Scott, *Women, Work, and Family* (New York: Holt, Rinehart and Winston, 1978). The family workshop persisted as Laura S. Strumingher shows in "The Artisan Family: Traditions and Transition in Nineteenth Century Lyon", *Journal of Family History* 2 (1977): 211–222.

20. Sewell,[17] offers well-chosen evidence for asserting of women that "their sex made them incapable — in the eyes of contemporaries and of the law — of exercising the paternal authority implied in a mastership," p. 31. Maité Albistur and Daniel Armogathe, *Histoire du féminisme française du môyen âge à nos jours* (Paris: Editions des femmes, 1977) discuss the question and cite evidence on both sides. The meticulously researched article of Natalie Zemon Davis, "Women in the Crafts in Sixteenth-Century Lyon," *Feminist Studies* 8, no. 1 (Spring 1982): 47–80, shows the wide range of women's work, documenting a few women's masterships in Lyon. Alice Clark found that women were seldom apprenticed in English guilds and that masters' wives could work along with their husbands only before the late seventeenth century. For a thorough discussion

see her *Working Life of Women in the Seventeenth Century* (New York: Harcourt, Brace & Howe, 1929). But for evidence of female apprentices in the silk industry see Sylvia Thrupp, *The Merchant Class of Medieval London* (Chicago: University of Chicago Press, 1980), p. 172; Eileen Power, *Medieval Women* (Cambridge: Cambridge University Press, 1976), p. 61; and A. Abram, "Women Traders in Medieval London," *Economic Journal* 26 (June 1916):276–285.

21.   Fagniez,[15] p. 16. Of 321 occupations listed in the tax rolls of 1292 and 1300, 108 were open to women and 80 were mixed; Albistur and Armogathe,[20] p. 35. See also Françoise Barrett, *Histoire du travail* (Paris: P.U.F., 1955), p. 35.

22.   Edmonde Charles-Roux *et al.*, *Les Femmes et le travail du moyen-âge à nos jours* ([Paris]: La Courtille, [1975]), pp. 33–35. Régine Pernoud, *La Femme au temps des cathédrales* (Paris: Stock, 1980) adds information from literary sources to the *Règlements sur les arts et métiers de Paris rédigés au xiiie siècle et connus sous le nom du "Livre des métiers"* d'Etienne Boileau, G.B. Depping, ed. (Paris, 1837).

23.   Hauser,[11] p. 152.

24.   Fagniez,[15] pp. 97–100.

25.   Davis,[20] p. 68. See also A. Rosembert, *La Veuve en droit canonique jusqu'au xiv siècle* (Paris: Dalloz, 1923), p. 145; and P. Boscou-Vance, "La condition des femmes en France et les progrès des idées féministees du xvie et au xviiie siècle," *L'Information Historique* 28, no. 4 (1966):139–141.

26.   For information on the silk industry at Lyon see Natalis Rondot, *L'Ancien régime du travail à Lyon* (Lyon: Alexandre Rey, 1897). Natalie Zemon Davis[20] sets these occurrences in a larger context, p. 68.

27.   Davis,[20] pp. 68–69, 71.

28.   Davis[20] and Fagniez,[15] p. 108.

29.   For information on the processes used in the silk brocade loom see Daryl M. Hafter, "From Mise-en-Carte to Loom in Philippe De LaSalle's Career," *Winterthur Portfolio* 12 (January 1977): 139–64. Women's work in fulling establishments is discussed in Walter Endrei, "Changements dans la productivité de l'industrie lanière au moyen âge," *Annales: Economies, Sociétés, Civilisations* 26 (Sept.–Dec. 1971): 1296.

30.   Fagniez,[15] pp. 108–09; The shift of occupation in the Italian woolen industry, as women entered the craft as weavers of plain goods is shown by Judith C. Brown and Jordan Goodman, "Women and Industry in Florence," *Journal of Economic History* 40 (1980):73–80.

31.   But see Liliane Mottu-Weber, "Apprentissages et économie genevoise au début du xviiie siècle," *Schweizerische Zeitschrift fur Geschichte* 20 (1970):340–347. She finds that few women had masterships in early eighteenth-century Geneva. For discussion of a cottage industry that had been taken over from men by the women see Reed Benhamou, "Verdigris and the Entrepreneuse," *Technology and Culture* 25, no. 2 (April 1984):171–181. These women protected their monopoly in the eighteenth century by ridiculing men who sought to enter the craft.

32.   For documentation of women's business activities in the Department of Le Nord before the women submerged themselves in religion and domesticity see Bonnie G. Smith, *Ladies of the Leisure Class: The Bourgeoises of Northern France in the Nineteenth Century* (Princeton, NJ: Princeton University Press, 1981).

33.   Levasseur,[18] vol. 2, p. 840. See Marilyn J. Boxer, "Women in Industrial Homework: The Flowermakers of Paris in the Belle Epoque," *French Historical Studies* 12, no. 3 (Spring 1982):401–423, to learn of the continuing success of this trade and its technological innovations in the nineteenth century. The independence of women in eighteenth-century women's and mixed guilds of Rouen — the hosiers, drygoods merchants and woolen cloth makers, fashionable plume makers, old and new clothes sellers, tailors, makers of gold braid, and seamstresses of religious vestments — is noteworthy. For a discussion of a mixed guild dominated by women see Daryl M. Hafter, "The

Spinners of Rouen Confront English Technology," *Proceedings of the International Conference on the Role of Women in the History of Science, Technology, and Medicine, Veszprém, Hungary, August 15-19, 1983,* 1 (Budapest 1983):70-75.

34.   Fifteenth-century clothiers of Paris declared that in their trade men and women were equal. J. Billioud, "Les classes industrielles en Provence, aux xive, xve et xvie siècles," in *Memoires de l'Institut historique de Provence* 6 (1929). Charles Schmidt remarked that the suppression of the "petit cadre corporatif" and the degrading of apprentice training removed protection from women and children workers. "Les débuts de l'industrie cotonnière en France 1760-1806," *Revue d'Histoire Économique et Sociale* 7 (1914):46.

Where Roof Meets Wall:
Structural Innovations and
Hammer-Beam Antecedents, 1150–1250

LYNN T. COURTENAY
2919 Oxford Road
Madison, Wisconsin 53705

IN AN AGE of mass-produced steel trusses, the structural mechanics of medieval timber roofs may be easily overlooked, especially when hidden by masonry vaults. Indeed, few may ask what kind of timber framing sustains those high gables above the vaults of the great cathedrals of Chartres, Bourges, or Amiens. How did the medieval carpenter cope with the essential problem of spanning space while ensuring both lateral and longitudinal stability against the forces of wind, gravity, and the limitations inherent in the medium of wood? These are basic concerns for any builder quite apart from aesthetic or social demands.

In an open-timber roof, where the structural carpentry is fully visible (FIGS. 1, A and B, and 2), the technological challenges of span and stability are augmented by a societal desire for artistically pleasing timberwork that harmonizes with the masonry structure which it surmounts. Thus open roofs, in contrast to those above vaults, are subject to the same interaction between style and structure that characterizes Gothic architecture in general. This dialectic between function and decoration is illustrated rather dramatically in hammer-beam construction, a roof type that developed exclusively in England in the fourteenth century, but some of whose structural antecedents were created in a quite different architectural context.

While the English exploitation of the hammer-beam form to create decorative roofs for great halls or angel canopies for parish churches is well known,[1] certain framing innovations that occurred much earlier in roofs hidden above masonry vaults and which laid the technological groundwork for this later development are considerably less familiar, nor have they been discussed in the context of the later English tradition. These innovations, chiefly in the area of the rafter foot where roof meets wall (FIG. 3), took place in the late twelfth and early thirteenth centuries in northern France and the Low Countries, both a period and a region characterized by vigorous experimentation

FIGURE 1A. London, Westminster Hall, 1394–1400. Interior looking south. (Photograph courtesy of F. Horlbeck)

in monumental carpentry and proto-industrial mechanization. It is the purpose of this paper to analyze from an English perspective the function of early bracket and cantilevered forms first developed in Continental roofs as a means of achieving stability in the new high structures of the early Gothic period.[2]

THE HAMMER-BEAM ROOF: DEFINITION AND STRUCTURE

All definitions of the hammer-beam roof agree that it is a roof type in which the principal frame contains a horizontal beam (the hammer beam) perpendicular to the wall and projecting at wall-plate level beyond the vertical plane of the masonry. Structurally the hammer beam is generally classified as an extended "sole-piece," the horizontal member making up the base of a rafter foot (FIGS. 4 and 5, A and B).

In the hammer-beam roof the hammer-beam frame is the main supporting truss (or series of trusses) for the common-rafters (FIG. 4), and these trusses form what are called "principals" or "principal rafters." In a typical early hammer-beam roof, such as Pilgrims' Hall at Winchester, ca. 1290 (FIG. 5), the hammer beam supports a vertical hammer post at its inner end; this post

FIGURE 1B. London, Westminster Hall, 1394–1400. Roof detail. (Photograph by W. Courtenay)

then rises to a longitudinal "collar-purlin," the structural member that ties the principals together lengthwise and helps carry the load of the roof. According to A. Emery's criteria, discussed in connection with the origins of the hammer-beam roof at Dartington Hall, the essential elements in a hammer-beam truss are: the hammer beam, the hammer post, and the square-set purlin, *viz.* a longitudinal purlin that is upright and not tilted to conform with the inclination of the rafters (*cf.* FIGS 4 and 5).[3] The lower bracket formed by the combination of the hammer beam, wall-post, and arch brace (not mentioned by Emery) is critical, however, in linking the timberwork to the masonry (FIGS.

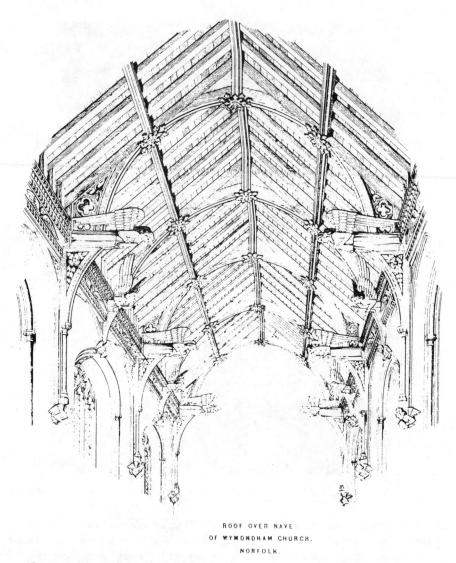

ROOF OVER NAVE
OF WYMONDHAM CHURCH.
NORFOLK.

FIGURE 2. Wymondham, Norfolk, angel roof of the Church of Saints Mary and Thomas of Canterbury (drawing: R. Brandon[7] p. 20).

1, 2, 5, and 6), and this brace is almost always present when the hammer-beam frame functions as a principal. Moreover, this lower bracket is not only a key stylistic element in hammer-beam roofs and an important area for decorative treatment (FIGS. 1 and 2), but also a chief functional component that

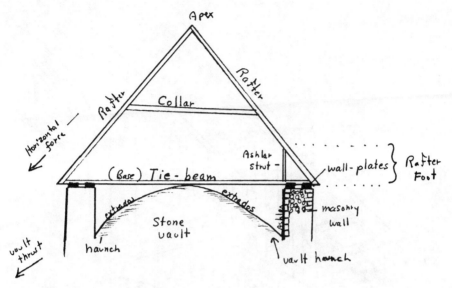

FIGURE 3. Collar-rafter roof above a vault.

emerged quite early in roof framing. Strictly speaking, however, it should be noted that this lower bracket structure occurs in roof types other than hammer beams, as for example beneath tie-beams where it historically appears to have originated.

STRUCTURAL EXPERIMENTATION IN ROOFS ABOVE VAULTS: THE RAFTER FOOT

While the general evolution of medieval roofs is a vast topic well beyond the scope of this paper, it should be noted, before concentrating on the specific configuration of the rafter foot, that aside from an upper tier of masonry flyers two major means were used by northern European carpenters from *ca.* 1150–1250 to control the lateral thrust of the roof at the top of the masonry walls: (1) massive tie-beams (FIG. 3) at the base of the rafter couples ("base-ties"), timber members spanning an entire vault or roof frame, which Fitchen has argued[4] were essential to the process of medieval masonry construction; and (2), the alteration of the pitch of the roof to an angle of 55 to 65 degrees.

Both of these solutions, however, created other problems to be solved by the carpenters. For example, base-ties in every frame of a large cathedral roof (the normal spacing for frames was usually about a meter or less) above a

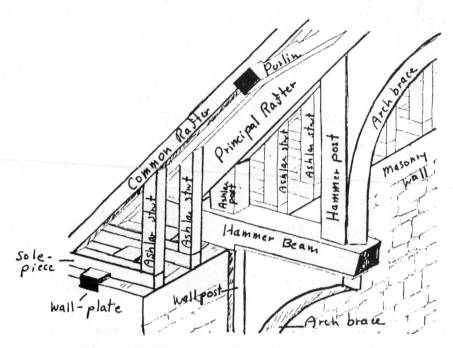

FIGURE 4. Hammer-beam framing (after Cordingley,[1] p. 86).

Gothic rib-vault meant that the walls supporting the roof had to be raised sufficiently for the base-ties to clear the extrados (external hump) of the vault (FIG. 3). This was naturally costly in material and labor to say nothing of the difficulty in finding suitable timbers for spans greater than 12 to 13 meters. While the steeper pitch of the roof gables (probably also desired for aesthetic reasons) accomplished a considerable reduction of horizontal thrust up to a point of inclination of 60 to 63 degrees,[5] this second means of adjusting for horizontal forces at the same time created a considerably larger surface area (and weight), somewhat like a sail, which might be subjected to increased wind forces and hence to greater leverage at the base of the roof structure. The insurance, therefore, of lateral and especially longitudinal stability against racking (lengthwise deformation of the structure) became even more critical issues as buildings and consequently roofs grew larger in height and span. Thus it is the creation of stability at the *base* of the roof that appears to have been one of the chief concerns of twelfth-century carpenters.

The response to these critical mechanical issues produced a number of innovations and frame designs to cope with the greater stresses of large-scale

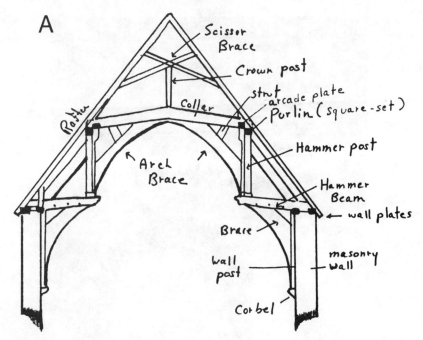

FIGURE 5A. Winchester, Hampshire, Pilgrims' Hall, *ca.* 1290. Transverse section of the roof (after J.T. Smith,[1] "Medieval Roofs: A Classification," p. 123).

structures and to accommodate the irregular space above a steeply pitched vault. Among the early and particularly significant developments dating to the twelfth century was the creation of a system of longitudinal bracing to secure tie-beams from turning, which might result from (1) a failure of the joints as the green timber dried, or possibly, (2) leverage caused by wind pressure. This was accomplished primarily by the insertion of single beams or pairs of timbers underneath the tie-beams along the top level of the masonry.[6] These members are called "wall-plates" (FIGS. 3 and 4), and the tie-beams are generally trenched to fit over them. Once established, wall-plates, either in pairs or as single timbers, became a constant feature of medieval roofs. They were used widely in the early Gothic period quite independently of base-ties, which during the thirteenth century were reduced in number or abandoned altogether as efficient trusses evolved and as more steeply pitched vaults intruded into the space occupied by the base-tie (FIG. 7). Moreover, the wall-plates performed a dual function: (1) to keep the tie-beams in place and (2) to furnish a "housing" (slot) for the ends of the rafters or for the horizontal base of a rafter foot known as a "sole-piece" (FIG. 8).

FIGURE 5B. Winchester, Hampshire, Pilgrims' Hall, *ca.* 1290. Detail of hammer-beam framing. (Photograph by W. Courtenay)

Concomitant with the development of wall-plates and sole-pieces as well as the carpenter's seeming preoccupation with the lower part of the roof's structure, was the creation of a simple but structurally important part of a roof known as a "rafter foot" (FIG. 3). This innovation was achieved by the addition of a short vertical post called an "ashlar strut" (aligned with the ashlar of the masonry, *i.e.*, the dressed stone used for the outer surface of the wall). This strut, in compression, braced the lower slope of the rafter and became the inner side of a triangle formed by the rafter and the horizontal sole-piece (FIG. 8B). Once developed the triangular rafter foot became one of the primary means of placing a timber roof on the *entire* width of the masonry wall.

The connection between this standard form of rafter foot and hammer-

FIGURE 6. Chichester, Sussex, Bishop's kitchen, *ca.* 1290. (Photograph by W. Courtenay)

beam construction has long been noted,[7] and a comparison of the horizontal hammer beam with the sole-piece is obvious (*cf.* FIGS. 4, 5, and 8B). What is important to keep in mind in relating these two structures is that both are part of a rigid triangle at the base of the roof frame, both extend generally across paired wall-plates from the outer plane of the wall into the interior of the building, and (as will become more apparent in this discussion) both are used as a type of cantilever. However, as the hammer beam evolved into a load-bearing truss supporting a purlin, its resemblance to the rafter foot became more formal than structural.

RAFTER FOOT VARIATIONS: EXTENDED SOLE-PIECES AND FLYING PLATES

Despite the relatively small number of surviving roofs in northern Europe from *ca.* 1150–1250, there appears to have been considerable experimentation in rafter-foot design. Behind this experimentation lay particular structural concerns of the medieval carpenter.

The simplest of these variations (other than changing the shape of the triangle by an alteration in roof pitch or by inclination of the ashlar strut) was to *enlarge* this triangular base by moving the ashlar strut and sole-piece inward so that they overhung the inner plane of the wall. This then appears

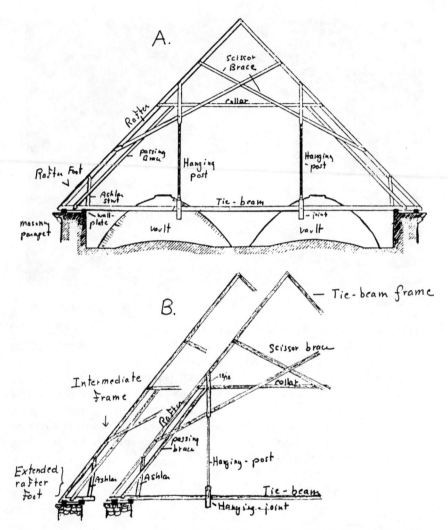

FIGURE 7. Angers, Musée St. Jean, Chapel Roof, *ca.* 1190–1200. A. Tie-beam frame (after Deneux,[20] p. 53). B. Detail of intermediate and tie-beam frames (after Ostendorf,[9] p. 15).

to be an important means by which carpenters used the geometry of their structure to adjust loading on the masonry and utilize the concept of positional gravity.[8] Such a construction was employed in the twelfth-century roof above the choir of the cathedral of Fritzlar,[9] and a particularly clear and rather remarkable example of an overhanging rafter foot exists in the collegiate church

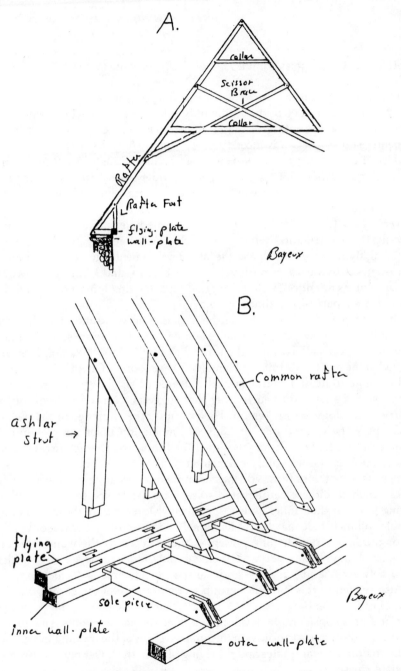

FIGURE 8. Bayeux Cathedral, N. Transept roof, *ca.* 1200. A. Scissor truss and extended rafter foot (after Deneux,[17] vol. II: D2558). B. Detail of rafter foot (after Deneux,[17] vol. II: D2561).

of St. Vincent at Soignies (Zinnik) in the old County of Hainault (17 km north of Mons).

The roof at Soignies (FIG. 9) is for our purposes an archetypal example. This roof has been dated to the period of Boudouin IV (1120–1171), who aided in financing the re-roofing of the tenth-century choir of St. Vincent.[10] The carpentry shows a very advanced type of structure for its date (probably third quarter of the century) with tie-beam frames alternating every fourth frame with enlarged rafter footing and, especially noteworthy, the use of longitudinal bracing by means of a wall-plate and an extra "flying plate" situated at the end of the extended sole-piece (FIG. 9). The rafter foot thus forms a cantilevered base for the lower part of the intermediate frames (those without base-ties). This structure, which thus locks the frames together by the longitudinal "flying plate," anticipates the later development of longitudinal bracing that occurs with the use of purlins (cf. FIGS. 4, 5, 14, and 17). Here, however, the carpenter's attention remains focused on achieving longitudinal stability at the lowest part of the timberwork.

The example of Soignies, while the earliest known to me of such a construction, is not, however, an isolated example. In fact, there are several remarkably similar roofs all dating to the thirteenth century and, like Soignies, also located in the ancient County of Flanders. These roofs at Ghent, Oudenaarde, and Damme (outside Bruges) exhibit a degree of similarity in construction to that of Soignies and to each other that suggests a certain "school" of carpentry technology in this region. For example, the churches of St. Nicolas in Ghent,[11] the Church of Our Lady of Pamele in Oudenaarde,[12] and the Church of Our Lady in Damme[13] (FIG. 10) all show the characteristic alternation of base-tie frames with weaker intermediate frames. Each of the lesser frames, as at Soignies (FIG. 9B), has prominent overhanging rafter footings, and in each case these cantilevered sole-pieces are joined to a longitudinal "flying plate" at their inner ends respectively. Moreover, all these roofs have double collars (Oudenaarde and Damme had king-posts), and the only form of longitudinal strengthening present is that offered by the wall-plates and the extra flying plate. The spans of Ghent and Oudenaarde are almost identical, between 7 and 8 meters, and the span of Damme is 10 meters.

Also related to this "Soignies group" is the framing of the choir of the cathedral of Tournai dating to 1243–1255.[14] This roof (span 12 meters), however, marks a considerable development over the smaller roofs, since it has longitudinal supports (purlins) in the upper section of the framing in addition to the customary wall-plates and the flying plate in the lower part of the intermediate trusses.

Finally, as an example of the continuation of the use of the cantilevered

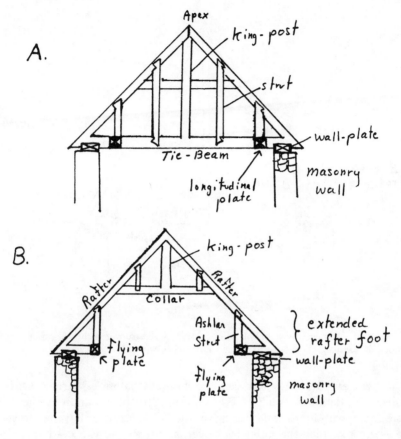

FIGURE 9. Soignies (Zinnik), Collegiate Church of St. Vincent, choir 1120–1171. A. Tie-beam frame (after Brigode,[10] p. 153). B. Intermediate frame (after Brigode,[10] p. 153).

sole-piece as a supporting member in intermediate frames, one sees later in the thirteenth century in the Reform Church at Oirschot in North Brabant, Holland, a striking form of an arch-brace and collar roof (FIG. 11).[15] Here, the principal frames with base-ties alternate with intermediate frames with extended rafter footings that support arch-braces, a type of so-called "false hammer-beam" construction developed somewhat later in English roofs, especially in parish churches (FIG. 2).[16]

While the greatest percentage of early roofs with extended rafter footings and flying plates is to be found in the Low Countries, there are also two surviving examples in northern France: the cathedral of Bayeux (FIG. 8) dating

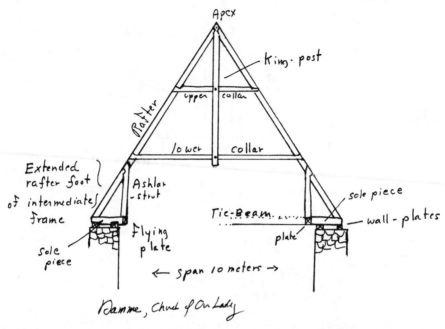

FIGURE 10. Damme, Church of Our Lady, *ca.* 1235. Reconstruction of the original intermediate framing of the main choir (after Janse and Devliegher,[10] p. 321).

to *ca.* 1210, and the fourteenth-century choir of Chalons-s-Marne.[17] One sees in Deneux's detailed sketch of Bayeux (FIG. 8B) the situation of the flying plate and how it serves as a housing for the ashlar strut. At Bayeux, however, the extent to which the rafter foot overhangs the wall is exceedingly slight in contrast to the group of churches seen in Flanders and associated with the framing at Soignies.

Taken as a whole, the group of roofs discussed above range chronologically from the mid-twelfth into the fourteenth century, and they suggest the presence of a strong interest in rafter-foot technology in northern France and especially the Low Countries in which roof frames with extended sole-pieces began to be locked together by longitudinal members in order to form a more stable and integrated structure that parallels the creation of the skeletal system of Gothic masonry. This constitutes a major development in the evolution of medieval roofs quite apart from their historical interest as resources for the later hammer-beam roof.

Finally, one example of the use of the extended sole-piece in a roof of considerable span and which illustrates the variety of experimentation in this period

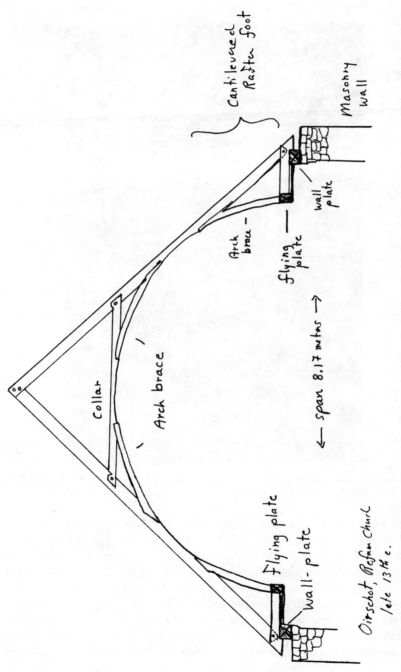

Cantilevered Rafter foot

Masonry wall

wall plate

Arch brace —

flying plate —

Collar

Arch brace

← span 8.17 metres →

Flying plate

wall-plate

Oirschot, Reform Church
late 13th c.

FIGURE 11. Oirschot, Brabant, Reform Church Roof, ca. 1290. Intermediate framing (after Janse and Devliegher,[10] p. 331).

FIGURE 12A. Great Coxwell, Oxon., Tithe Barn of Beaulieu, *ca.* 1300? Intermediate framing of the aisle. (Photograph by W. Courtenay)

is the roof above the chapel of the Hospital of St. John at Angers (Musée St. Jean) dating to the end of the twelfth century and having the large span of 18 meters. While this roof (FIG. 7) does not utilize a flying plate in conjunction with the extended sole-piece, the framing design illustrates another important category of roofs that employed extra parallel braces (a "passing brace")[18] to achieve greater strength in what is a strikingly light-weight construction in comparison to the total dimensions of the structure. As in the Soignies roof, the tie-beam frames at Angers are reduced in number and actually rest on the masonry of the vaults (FIG. 7A), which here are the high domical type

FIGURE 12B. Great Coxwell, Oxon., Tithe Barn of Beaulieu, *ca.* 1300? Detail of an original "proto-hammer beam." (Photograph by W. Courtenay)

developed in Angevin building of this period. The rafter-foot construction shown in the intermediate frames at Angers exhibits the familiar inward-projecting sole-piece, which in this case supports the parallel rafter, or passing brace and the ashlar strut (FIG. 7B). This configuration of a projecting sole-piece sustaining parallel timbers (*i.e.*, the rafter and the passing brace) has remarkably survived in an English example from the early fourteenth century, *viz.* at Great Coxwell (FIG. 12)[19] and is seen somewhat earlier in the intermediate frames of the roof above the vaults at Notre Dame at Etampes.[20] The relationship among the rafter footings and passing brace construction

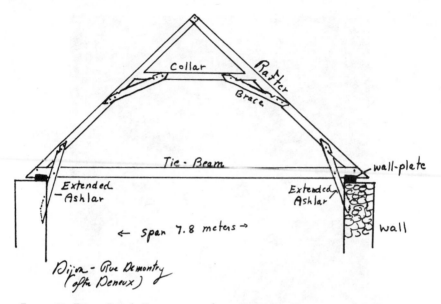

FIGURE 13. Dijon, Rue de Demontry, roof, *ca.* 1190 (after Deneux,[17] vol. I: D6940).

at Angers, Great Coxwell, and Etampes illustrates the general diffusion of carpentry techniques at this time, a phenomenon no doubt enhanced by the growth of monastic orders in the late twelfth century.

RAFTER FOOTING WITH EXTENDED ASHLAR STRUTS OR WALL POSTS

In addition to enlarging the triangular rafter foot by using the sole-piece as a cantilever in intermediate frames at a time when base-ties were being reduced in number, there is also evidence that carpenters sought to achieve greater stability against lateral thrust by using a T-brace or "hook" principle of construction. This involved the concept of substituting a different design in place of the base-tie, and carrying the timber framing down the inner plane of the wall where it could be supported by the masonry.

The earliest example of this kind of construction known to me occurs in a late twelfth-century roof in a building in Dijon in the rue Demontry (FIG. 13).[21] As in the ecclesiastical roofs previously discussed, the tie-beams have been reduced in number, here to every sixth frame, and as is common for roofs of this period in France, every frame is independent from the others with no longitudinal linkage. The ashlar strut, fastened to the rafter by a double-

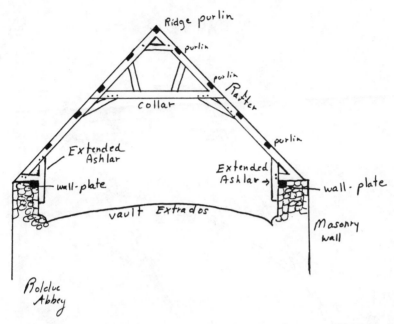

FIGURE 14. Kerkrade, Holland, Rolduc Abbey Church, transept framing twelfth century (after Janse and Devliegher,[10] p. 313).

pinned lap joint, is extended down past the sole-piece or base-tie to which it is also pegged, so that its end is embedded in the wall. The idea of the carpenter seems clearly to stabilize the roof frame by anchoring it to the masonry, and in so doing he has created both a brace against lateral sliding on the top of the wall and by inserting the ashlars into the stone-work has also protected the frames against turning and hence racking.

A type of construction related to this domestic example in Dijon but whose date is uncertain (part of it certainly late medieval) is the alleged twelfth-century roof above the vaults at Rolduc Abbey in Kerkrade, Holland, also in the ancient County of Flanders.[22] At Rolduc (FIG. 14) the wall-plate is placed in a slot in the masonry (something which also occurs in late medieval English roofs)[23] and the ashlar strut is extended downward and becomes a wall-post. This construction forms a kind of "notch" over the wall, and its design is evidently to prevent lateral spreading at the base of the roof.[24] A very similar construction can also be seen in a later open-roof at Tedburn St. Mary, Devon.[25] While the framing designs at Dijon, Rolduc and Tedburn St. Mary are different from one another, the essential idea of extending the ashlar strut downward

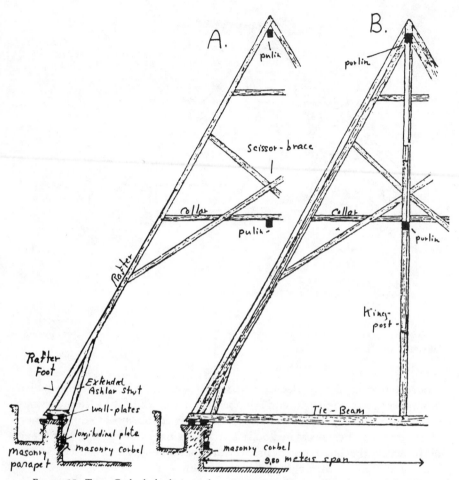

FIGURE 15. Tours Cathedral, choir roof, ca. 1250–1300. Intermediate framing (after Osten-dorf,[9] p. 19).

and linking the rafter foot structurally to the masonry is fundamental in all three examples.

By the late thirteenth century the structural concept involved at Dijon and Rolduc appears in a large cathedral roof, that above the choir of Tours. In the framing at Tours ca. 1250–1300 (FIG. 15), the ashlar strut is extended down the wall and is received by: (1) a wall-plate and (2) a masonry corbel immediately beneath the wall-plate. Because the roof is so steeply pitched (64 degrees), the carpenter has apparently used the ashlar strut and masonry corbel as a means of rafter footing specifically designed to transfer part of the vertical load of the roof frame to a point lower down the wall on its inner side. Also it should be remembered that the wall-plate, here placed far down the wall, is still a means of longitudinal bracing and is analogous, I think, to the extra flying plates seen earlier in roofs associated with Soignies, since this longitudinal member gives support where loading is occurring at the extreme base of the frame.

From the above examples involving the extended ashlar strut, which resembles the wall-post found in later hammer-beam construction, two important principles emerge: (1) the idea of transferring a portion of the roof's weight down the masonry, especially where, as at Tours, the parapet upon which the roof rests in exceedingly thin (FIG. 15); and (2) the creation of a brace as at Dijon and Rolduc against outward spreading by using the wall and timber *together*.

THE SADDLE BRACKET

While the triangulated rafter foot in its various forms remained a standard component of medieval roof construction, a specialized type of beam, termed by Ostendorf a "saddle" (*Sattelholz*),[26] emerged at the beginning of the thirteenth century in conjunction with experimentation in large-scale roof framing at Paris and nearby Mantes and Meaux, and slightly later in Flanders. This new technique involved the use of a short horizontal beam, the "saddle," beneath the ends of a base-tie to which it was secured by wooden pegs. The saddle was then supported below by a post and a slanted brace (FIGS. 16–19), thus forming what I shall call a "saddle bracket." According to a recent discussion of this device by John Smith,[27] the saddle emerged primarily as a means of preventing tie-beams from turning when joints became less reliable as the timber shrank from aging and drying (most medieval carpenters were obliged to use green timber). But, at the same time, it was realized that the saddle in conjunction with a brace, which so conveniently fitted the triangular space between the vault extrados and the tie (FIG. 3), could also serve the purpose of transferring part of the weight of the roof to a point of loading further down the wall. Clearly, the idea of the carpenter was to secure as great a stability as possible at the base of these new wide and steeply pitched roofs, which in high wind conditions would be subject to leverage at their base. Moreover, base-ties were probably critical platform supports during vault erection. As the extant examples illustrate (and it is important historically that they first appeared at Paris where other major architectural innovations also occurred),[28] the carpentry method chosen to solve this evident concern for stability created

FIGURE 16A. Paris, Notre Dame, choir roof, *ca*. 1220. Choir saddle bracket (upper part modern). (Photograph by L. Courtenay)

FIGURE 16B. Saddle brackets with tie-beams and hanging post. (Photograph by L. Courtenay)

a greater integration between the carpentry and the masonry structure, a fact that deserves serious attention.

The earliest known use of this device occurs in the choir of Notre Dame at Paris *ca.* 1200, but whose framing was rebuilt around 1220 using some of the twelfth-century timbers.[29] The choir roof (FIG. 16), described in some detail by Viollet-le-Duc,[30] consists of closely spaced frames of light scantling (section) and contains large saddles generally with double braces. In each example of the choir saddle brackets, the wall-post is set slightly into the masonry and rests on a small masonry corbel.

This use of a bracket support in the lower part of the roof frame evidently found favor, since the same device with modifications was repeated in the framing of the nave of Notre Dame *ca.* 1230. In the nave (FIG. 17, A and B), which shows considerable technological advance over the framing of the choir, both saddle brackets and longitudinal supports are used. The supports beneath the saddles in the nave of Notre Dame (FIG. 17B) constitute a slightly inclined wall-post from which spring lateral braces connected to a longitudinal plate. A diagonal brace then connects the saddle to the wall-post, and the

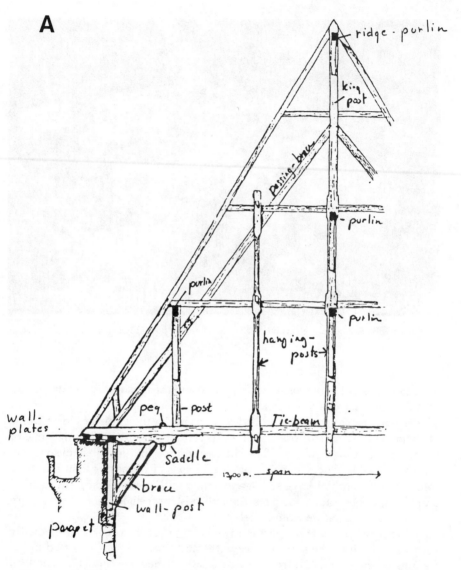

FIGURE 17A. Paris, Notre Dame, nave framing, *ca.* 1230. Framing of the nave, transverse section (after Ostendorf,[9] p. 20).

whole of this lower bracket rests on some kind of masonry support that varies in configuration from bay to bay. Moreover, and especially important to the structure, a vertical post carrying a lengthwise brace (or purlin) is positioned

FIGURE 17B. Paris, Notre Dame, nave framing, *ca.* 1230. Nave framing, detail of saddle bracket and "purlin-post." (Photograph by L. Courtenay)

at the end of the saddle bracket. Since this vertical strut or "purlin post" is tenoned into the tie-beam and is not a hanging post (in contrast to the large "king" and "queen" posts, which have tension joints), the idea of the carpenter seems clearly to utilize the saddle bracket to transfer a portion of the load carried by the "purlin post" to a point lower down the masonry wall. The structural configuration of the saddle bracket, which has not been commented on, supports this interpretation. Although the longitudinal purlin supported by the vertical post is not fully integrated with the main outer rafter, I think that in the nave of Notre Dame the saddle bracket was conceived in conjunction with the "purlin post" not only as a support for the ends of the tie-beam and as a guard against the turning of the base-tie but also as a compressive strut designed to take a portion of the roof's weight down onto the wall where it thickens at the upper edge of the vaults.

A saddle is also used beneath the ties in the king-post roof at Mantes,[31] which dates to the early thirteenth century. Here the bracing of the tie is far

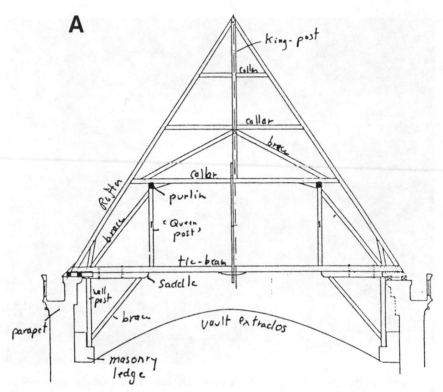

FIGURE 18A. Meaux Cathedral, choir roof, *ca.* 1210. Transverse section of the choir (after Deneux,[17] vol. II: D6985).

less substantial and is not associated with the vertical post and purlin as in the nave of Notre Dame at Paris. At Mantes, it would appear that the saddle construction was used as an insurance against the turning of the base-tie and was not intended to redistribute any of the thrust or loading of the roof.

Among the early examples of saddle structures, the roofs above the nave, choir and transepts of the cathedral at Meaux (45 km east of Paris), roughly contemporary with the framing of the choir of Notre Dame, exhibit well-developed forms of this prominent bracket structure. At Meaux (FIG. 18, A and B) the saddle, wall-post, and diagonal brace form a large bracket that extends approximately a quarter of the span (12 meters) of the roof. Near the end of the saddle (but not consistently positioned throughout the structure) a vertical post, resembling in this case a kind of "aisle post," is tenoned into the base tie (FIG. 18A). As in the nave of Notre Dame, this post carries a longitudinal support (purlin).[32] Beneath the saddles at Meaux, the wall-post

FIGURE 18B. Meaux Cathedral, choir roof, *ca.* 1210. Meaux, south transept, saddle bracket. (Photograph by L. Courtenay)

and brace rest upon a purposely built masonry ledge similar in construction to the corbeled wall-posts in roofs of the nave and choir at Notre Dame. The substantial scale of the saddle bracket in a roof of very light construction and the consistent use of a masonry support beneath these structures, seems to

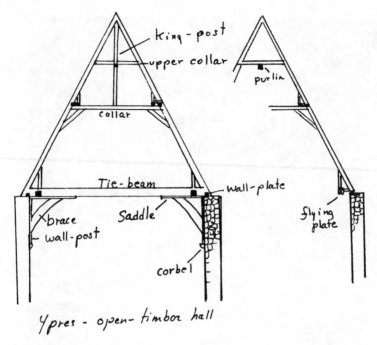

FIGURE 19. Ypres, Market Hall. Framing of the east wing, *ca.* 1250 (after Janse and Devliegher,[10] p. 360).

leave little doubt that the carpenter and his colleague, the master mason, intended the combined forms of "purlin post," saddle bracket, and masonry support to form a compressive strut to transfer a portion of the roof's load lower down the masonry, *viz.* to the inner side of the main supporting pier from which the parapet extends.

From this evidence, and the continued use of saddle brackets at a later date, one can conclude that in roofs of large span in which a base-tie was retained to counteract horizontal thrust, saddle brackets in conjunction with longitudinal supports and hanging posts ("queen" and "king" posts) represent an important early thirteenth-century solution to problems of construction, span, and loading of a large cathedral roof. Thus the framing design of the timberwork contributes in no small measure to the general success of the Gothic skeletal structure, where precision in handling weight and thrust is critical. Paris, followed by Meaux, seemingly laid the foundations of a framing technology that was applied to structures of large span in the later Middle Ages, as for example in the Chapter House at York Minster (span 17.4 meters)[33] dating to the end of the thirteenth century, where one also finds saddle brackets beneath base-ties and the use of large hanging or tension posts (central kingpost and queen-posts radially disposed).

While all the above examples of saddle structures are found in roofs above vaults and were perhaps developed to utilize efficiently the triangular space between the extrados of the vault and the masonry wall (*e.g.* FIGS. 3, 16–18), braced saddles also appeared in *open-roofed* halls in Flanders in the middle of the thirteenth century and formed a lower stage of a multi-tiered construction, a development perhaps influential for both hammer-beam and tie-beam roofs appearing slightly later on the other side of the Channel. For example, the market hall at Ypres (FIG. 19) has prominent saddle brackets with decorative carving which support massive tie-beams. Moreover, this type of construction continued in domestic and civil architecture in Holland into the fourteenth and fifteenth centuries. It is not surprising that the coastal regions of the Channel should share certain technological resources. The use of a bracket and wall-post to gain stability at the base of a roof frame and to carry a load down the wall is one of the principal structural functions of the hammer beam as it initially developed. This can be observed, for example, in the thirteenth-century kitchen of the Bishop's Palace at Chichester (FIG. 6) and in the roof of Pilgrims' Hall at Winchester (FIG. 5). Both of these first-known English examples of hammer-beam construction (*ca.* 1290) and those which appear only slightly later are located geographically in southeastern England, an area particularly influenced by Continental carpentry techniques. Thus, there seems good reason to believe that the use of a compressive brace beneath a saddle, likely initiated at Paris and later spreading to open-roofs in Holland, had an influence on carpentry technology in England. Moreover, although the saddle bracket was created for purely functional reasons and in the context of a particular kind of architectural space, this type of construction illustrates the close relationship between masonry and carpentry that was to become both structurally and later aesthetically such a prominent feature of hammer-beam roofs in England.

VILLARD DE HONNECOURT

As a conclusion to the foregoing discussion of the innovations that anticipate some of the structural and formal elements of hammer-beam design, it is pertinent historically to consider the drawings of timber roofs contained in the famous "Sketchbook" of Villard de Honnecourt, which dates to the second quarter of the thirteenth century, only a generation after bracket structures first appeared in northern French roofs. Villard's roof designs are seldom dis-

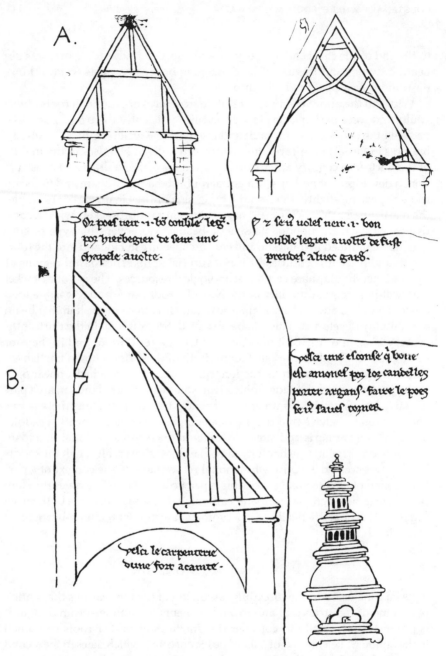

A.

B.

Or poef neir .i. bõ couble leg̃
por hierbegier de seur une
chapele auotre.

Fz z seū uolef neir .i. bon
couble legier auotre de fuſt
prendes aluec garß.

Vesci une efconfe q̃ boue
eſt amonef por los candelles
porter arganſ. faire le poeș
feū fauef comer

Vesci le carpenterie
dune fort acamte.

FIGURE 20. Villard de Honnecourt, Roof design, *ca.* 1235 (Reproduced from Lassus,[36] pl. 27).
A. "Hammer-beam" type frame. B. Aisle roof frame.

118

cussed in detail, and there is a general lack of clarity in the understanding of these drawings vis-à-vis the hammer beam and the extant structures of this period.

Regardless of exactly what sort of personnage one considers Villard to have been,[34] it has been well established that the "Sketchbook" (Paris, Bibl. Nat., Ms. Fr. 19193) is for the most part an autograph work dating to *ca*. 1235–40. Among the numerous subjects in which Villard was interested, structural carpentry, mentioned in the author's introduction, is certainly one of the most important, and it is unfortunate, as Willis notes, that four leaves of this portion of the manuscript have been lost.[35] However, the surviving folios relating to the craft of carpentry, which include a number of ingenious machines and mechanical devices, illustrate several types of roof frames, two of which pertain particularly to the hammer-beam antecedants discussed in this essay.

The first example (FIG. 20A) reproduced from the facsimile edition of Lassus, plate 33, the verso of the seventeenth leaf of the manuscript, shows a steeply pitched rafter couple braced mid-way by a collar surmounted by a vertical post which extends to the roof's ridge, a "king-post" (*cf*. FIGS. 9 and 10). The lower portion of the frame illustrates a greatly enlarged rafter foot composed of a horizontal beam that rests directly on the wall *without* wall-plates and extends inward beyond the vertical plane of the wall in a manner similar to the rafter footings in the intermediate frames at Soignies. This horizontal sole-piece, which one may regard as a proto-hammer beam, forms in Villard's frame a cantilever which is joined to the rafter at its outer end and supports a strut for the rafter on its inner end. This is not a hammer-beam roof, but rather an enlarged rafter foot whose purpose is to stabilize the frame. There is no longitudinal purlin indicated in the drawing, and as is clearly evident, the wall is thick, the span small, the projection of the footing slight, and thus there is no need for a brace or wall-post beneath the cantilevered beam or any redistribution of loading.

The text that accompanies this drawing instructs the reader that: "This is a good, light (or easily wrought) roof for the wall of a vaulted chapel." In the drawing, the vault is indicated in plan below by a hemicycle segmented by six ribs converging at an arbitrary center. This certainly appears to be a design for an apsidial or ambulatory chapel and resembles, except for the number of ribs, Villard's sketch of the ground plan of the cathedral of St. Etienne at Meaux.[36] It is important to note that the roof design in question is meant for a specific location, *viz*. a chapel, and therefore, the architectural scale and especially the span for which Villard's roof is intended is relatively small in comparison to the roofs above vaults surveyed in this study, most of which exceed 10 meters. Also, it is likely that the roof Villard has drawn is in fact a single frame designed to obviate a tie-beam at the crown of the vault of the semicircular chapel. This solution, similar to that developed in twelfth-century scissor trusses, thus allows the extrados of the vault to rise above the masonry walls.

In relationship to what actually existed in roof construction in the region of northern France and southern Belgium which Villard might have known, Villard's small-scale frame is structurally less impressive than those Villard himself might well have seen, *i.e.*, with wall-plates and other means of longitudinal bracing. Villard's so-called "hammer-beam roof" is certainly not new, but reflects the well-established practice for some forty years of using a cantilevered rafter foot as a triangular brace in the lower part of the weaker intermediate frames of early Gothic roofs.

On the same folio of Villard's manuscript, there is a second drawing (FIG. 20B) also illustrating the inward-projecting sole-piece. This occurs in a rather curious design for a pent roof of an aisle. The text accompanying the drawing says: "Here is the frame of a strong pent roof (*dune fort acainte*)." In this frame the horizontal sole-piece extends approximately half the span of the aisle (the vault is represented by the hemicycle below). From this cantilevered beam rise three struts whose structural functions are ambiguous. The outer two are similar to ashlar struts or struts from tie-beams found in Romanesque roofs, and they are presumably braces for the lower part of the rafter. The third and innermost post would also appear to be a brace supporting the mid-point of the rafter, but it is unclear how this operates. The understanding of this post is contingent upon how one interprets the rectangular block that appears in Villard's sketch at the end of the cantilevered beam and directly beneath the innermost post. The "block" might be interpreted as an extension of the post and thus be a representation of a tension joint, such as those that were utilized in hanging-post construction at Angers and Paris (FIGS. 7 and 17); however, this is unlikely, since it is not supported sufficiently at its apex. Moreover, this post is joined to the rafter at a point along the slope where the rafter is loaded and subject to sagging, so that a hanging-post at this point would be structurally counter productive.

A second and more viable interpretation, which was suggested by Willis but not fully explained,[37] is that the "block" beneath the innermost post represents a longitudinal member, in short, a type of "flying plate" or lengthwise brace at the base of the roof. Since this longitudinal plate occurs beneath the cantilevered beam, it must be supported elsewhere, perhaps by transverse ties in other frames or, as Willis suggests, by the masonry vaults.[38] The longitudinal timber would then support the extended sole-piece and change its function from a cantilever to a beam supported on both its ends, which in fact would be, as Villard says, a strong construction. This presumed longitudinal plate of Villard's is similar in its placement to the flying plates which were well established at Soignies, close to Cambrai where Villard lived, and which were evidently widely used in Flanders.

These brief observations about Villard's knowledge of structural carpentry indicate that in all probability Villard did draw roof frames which he actually saw, and this did *not* include a fully framed hammer-beam roof as it is normally defined. His drawings, ambiguous in their rendering of structural situations, manifest a limited understanding of various designs which were mechanically quite adequate for the small spans with which he was concerned. But, his lack of structural precision further substantiates the view that Villard was not himself a builder. A comparison of Villard's first sketch, which is a simple collar-rafter roof with a king-post extending from the collar to the apex (FIG. 20A), with the intermediate framing of the roof at Soignies (FIG. 9B), which also has a simple collar, king-post, and extended rafter foot, only underlines the fact that Villard's sketches are an important, although somewhat later, historical witness to the experimentation in roofing technology *ca.* 1180–1250, especially for Flanders, even if the drawings in question do not measure up to the level of structural sophistication already seen in roofs of large span.

SUMMARY

The examination of several technological innovations developed chiefly to improve longitudinal stability in large-scale roofs above masonry vaults has revealed considerable experimentation in the use of cantilevered or bracket structures (saddles or wall-posts) placed at the base of the roof frame in order to handle or redistribute loading. Several of the early developments discussed relate to the structure of the hammer-beam roof: (1) the cantilevered rafter foot; (2) the ashlar strut extended downward; and (3) the saddle bracket. These forms, as seen in the extant examples especially in the regions of Paris and Flanders, illustrate that carpenters of the early thirteenth century understood empirically the different functions of a bracket structure in compression, which transferred a portion of the roof's weight further down the masonry wall, in contrast to a rafter foot with an ashlar strut extended to form a wall-post in tension (pendant post) whose function was similar to that of a tie-beam: to prevent outward movement of the roof on the top of the wall. Several of these techniques, which were developed in the previous generation, are recorded in the famous *Sketchbook* of Villard de Honnecourt. But the important growth in structural mechanics appears to have occurred in roofs whose larger dimensions presented new challenges for carpenters of *ca.* 1200. By the late thir-

teenth century, these purely structural elements, which might also have facilitated actual construction and which were efficiently adapted to the enclosed
area above vaults, emerged in a new *spatial* context: the un-aisled open-timber
hall, such as that at Ypres. Considering the development of open-timber roofs
in England, the well-known economic ties between Flanders and southeast
England might have provided an avenue for the English importation of the
kinds of technological innovation examined in this essay.

Factors other than pure mechanical capability, however, contribute to the
appearance and exploitation of a particular architectural form. This is especially true of the great flowering of the hammer-beam roof in England, where
this domestic roof type (appearing *ca.* 1290 and chosen no doubt for its decorative potentiality) evolved in response to general societal preferences for large
un-aisled halls, where a cantilevered bracket linked to the masonry wall and
supporting a vertical post (the hammer post) expedited the achievement of
cultural and aesthetic desires. Although this study, purposefully limited to
Continental early Gothic structural innovations that anticipate hammer-beam
construction, brings us only to the threshold of that subsequent development
in open roofs, the techniques available by 1250 formed a rich repertory from
which later carpenters could choose solutions to new problems. The experiments of the late twelfth century were eventually grafted to the needs of a
later age with the same creativity as the carpenters who first evolved these
structures to meet specific challenges presented by the new scale of Gothic
masonry.

NOTES AND REFERENCES

1. F.H. Crossley, *Timber Building in England* (London: Batsford Ltd., 1951). J.T. Smith, "Medieval Aisled Halls and Their Derivatives," *Archaeological Journal* 122(1955): 76–94. J.T. Smith,
"Medieval Roofs: A Classification," *Archaeological Journal* cxv(1958): 111–149. J.B.L. Tolhurst,
"The Hammer-beam Figures of the Nave Roof of St. Mary's Church, Bury St. Edmunds," *Journal
of the British Archaeological Association*, Series 3, 25(1962): 66–70. R.A. Cordingley, "British
Historical Roof-types and Their Members: A Classification," *Transactions of the Ancient Monuments Society* n.s. 9(1961): 73–117.
2. This category of Continental roofs excludes cruck construction and the stave churches of
Scandinavia and is in reference to the widespread European carpentry tradition using post and
beam construction and timber frames of general uniformity in design and scantling (section of
the timbers).
3. A. Emery, *Dartington Hall* (Oxford: Oxford University Press, 1970) p. 237. *Cf.* also the important discussion of John Smith on the evolution of the hammer beam in relationship to aisled
construction. Smith,[1] "Medieval Aisled Halls," pp. 87–90. *Cf.* J. Crook, "Pilgrims' Hall, Winchester." In *Proceedings of the Hampshire Field Club Archaeological Society* 38(1982): 85–101;
and J. Munby, "The Date of the Pilgrims' Hall, Winchester," in the same journal, 40(1984): 130–133.

4. J. Fitchen, *The Construction of Gothic Cathedrals* (Chicago and London: University of Chicago Press, 1981) p. 26.

5. I am grateful to Prof. Robert Mark for a discussion of the mathematical formula to determine the relationship between roof pitch and horizontal force.

6. J.T. Smith, "Mittelaltesliche Dachkonstructionen in Nordwesteuropa." In *Fruhe Holzkirchen in nordlichen Europa* (Hamburg: Helms Museum, 1981) pp. 382–383. I am particularly grateful to John Smith for providing me with a typescript of his manuscript.

7. R. Brandon and J. Arthur Brandon, *The Open Timber Roofs of The Middle Ages* (London: David Bogue, 1849), p. 22.

8. *Cf.* M. Clagget, *The Science of Mechanics in the Middle Ages* (Madison: University of Wisconsin Press, 1959) pp. 69–112.

9. F. Ostendorf, *Geschichte des Dachwerks*, (Leipzig and Berlin: B.G. Teubner, 1908) p. 9.

10. S. Brigode, "L'Architecture Religieuse dans le Sud-Ouest de la Belgique," *Bulletin de la Commission Royal des Monuments et des Sites* I (1949): 150. *Cf.* H. Janse and L. Devliegher, "Middeleeuse Bekappingen in Het Vroegere Graafschap Vlaanderen," *Bulletin de la Commission Royale des Monuments et des Sites* 13(1962): 318–319.

11. Janse and Devliegher,[10] p. 319.

12. *Ibid.*, p. 320.

13. *Ibid.*, p. 321.

14. *Ibid.*, p. 339.

15. *Ibid.*, pp. 331–333.

16. Cordingley,[1] pp. 89 and 105.

17. H. Deneux, *Charpentes* (Paris: Centre de Recherches sur les Monuments Historiques, 1959–60) vol. II: D2558 and D25561. *Cf.* Ostendorf,[9] p. 80.

18. J.T. Smith, "The Early Development of Timber Buildings: The Passing-brace and Reversed Assembly." *Archaeological Journal* 131(1974): 238–263.

19. W. Horn and E. Born. *The Barns of Beaulieu at Its Granges of Great Coxwell and Beaulieu St. Leonards* (Berkeley and Los Angeles: University of California Press, 1965) p. 17. See also W. Horn, "The Potential and Limitations of Radiocarbon Dating in the Middle Ages: An Art Historian's View." In *Scientific Methods in Medieval Archaeology* (Berkeley, Los Angeles and London: University of California Press, 1970) pp. 23–87. Julian Munby has kindly informed me that the barn of Great Coxwell has been most recently dated by dendrochronology to *ca.* 1300–1310.

20. Ostendorf,[9] p. 18. The choir of Etampes has been dated by Deneux to the first half of the thirteenth century. H. Deneux. "L'Evolution des Charpentes du XI^{me}au XVme Siècles," *L'Architecte.* IV(1927): 59.

21. Deneux,[17] vol. I: D6940. Deneux,[20] p. 51.

22. Janse and Devliegher,[10] p. 313.

23. F.E. Howard, "On the Construction of Medieval Roofs," *Archaeological Journal* 71(1914): 316.

24. This term was used by Prof. J. Heyman in his structural analysis of the framing of Westminster Hall: J. Heyman, "Westminster Hall Roof," *Proceedings of the Institute of Civil Engineers* 37(1967): 153 and 160.

25. Howard,[23] p. 325, Fig. 16.

26. Ostendorf,[9] pp. 20 and 93 especially.

27. Smith,[6] p. 383.

28. W.W. Clark and R. Mark. "The First Flying Buttresses: A New Reconstruction of the Nave of Notre Dame de Paris," *Art Bulletin* 66(March 1984): 47–65.

29. I am indebted to the staff of the Monuments Historiques de France for making certain drawings

accessible to me and particularly to M. Françoise Commenges, Architect of Notre Dame, for making it possible for me to study and photograph the roof structure. Deneux,[20] p. 57.

30. E. Viollet-Le-Duc, *Dictionnaire Raisonné de L'Architecture* (Paris: E. Martinet, 1875, vol. 3, pp. 12–16, esp. p. 13. Fig. 9; *Cf.* also Deneux,[20] p. 57.

31. Deneux,[20] p. 58, Fig. 88.

32. It should be noted that this post is not so closely associated with the saddle bracket as at Notre Dame, since at Meaux the placement of the "purlin post" varies considerably from bay to bay and in different sections of the framing of nave, choir and transepts; thus the transverse section shown in the single frame by Deneux (FIG. 18 A in my text) is somewhat misleading.

33. J.Q. Hughes, "The Timber Roofs of York Minster." *Yorkshire Archaeological Journal* **XXXVIII**(1955): 474–495; C.A. Hewett, *English Cathedral Carpentry* (London: Wayland Publishers, 1974) pp. 74–77; and G.E. Aylmer and R. Cant, eds. *A History of York Minster* (Oxford: Clarendon Press, 1977), p. 138.

34. C. Barnes, *Villard de Honnecourt the Artist and His Drawings, a Critical Bibliography* (Boston, MA: G.K. Hall and Co., 1982).

35. R. Willis, trans. and ed., *Facsimile of the Sketch-book of Wilars De Honnecourt* (London: John Henry and James Parker, 1859) p. 102.

36. J.B.A. Lassus, ed, *Album De Villard De Honnecourt* (Paris: Imprimerie Imperiale, 1858) pp. 121–124, pl. 27.

37. Willis,[35] p. 106.

38. *Ibid.* It is interesting to note that above the extrados of the vaults in the north transept of the Cistercian abbey of Pontigny (Yonne, France) the mason built a rectangular support that projects from the crown of the vault. A longitudinal timber beam rests on this masonry block and is used to support the mid-point of the tie-beam, which has been moved from the base of the roof frame to a higher position. Deneux dates this construction to *ca.* 1200, exactly the period when the interaction between carpentry and masonry structures becomes more important. Pontigny is illustrated in Deneux,[17] vol. I: D6919.

High Gothic Structural Development: The Pinnacles of Reims Cathedral

ROBERT MARK and HUANG YUN-SHENG

School of Architecture
Princeton University
Princeton, New Jersey 08544

HISTORICAL BACKGROUND

THE NATURE of the technical resources employed to develop the High Gothic cathedral with its new, lighter structural systems during the second half of the twelfth century has been one of the more persistent puzzles in the history of architecture. Recently, however, new insights into design and construction methods of the era have been afforded by modern structural studies of extant monuments. These have indicated, for example, that the medieval builder used an experimental design method that embraced the actual building: modifications made to the structure to eliminate tensile cracking observed in newly set, weak lime mortar with the removal of construction scaffolding were probably an important source of structural innovation.[1]

The studies indicated also how an earlier building could have acted as an approximate "model" to confirm the stability of a new design. But this approach may not have been adequate for the planning of what appears to be the key building for Gothic structural devlopment, Notre-Dame de Paris (begun *ca.* 1160). The relatively lightly constructed central vessel of the new cathedral was a full eight meters taller than its highest Gothic predecessor. And since wind speeds at higher elevations are greater and wind pressure is proportional to the square of the speed, earlier experience with lower-profiled, more heavily massed churches could not have fully prepared its builders to cope with the new environment. Owing to massive reconstructions made to the cathedral after 1225, however, just how the design problem was surmounted in the original Gothic construction at Paris remained unclear.

From archaeological clues still preserved in Notre-Dame and in contemporaneous buildings from the Paris region as well as from drawings and photographs made before further mid-nineteenth-century restoration, a new reconstruction was determined of the original structural configuration of the nave (FIG. 1a).[2] Evidence was found in the archaeology of the first examples of medieval flying buttresses, which were probably erected as an emergency solu-

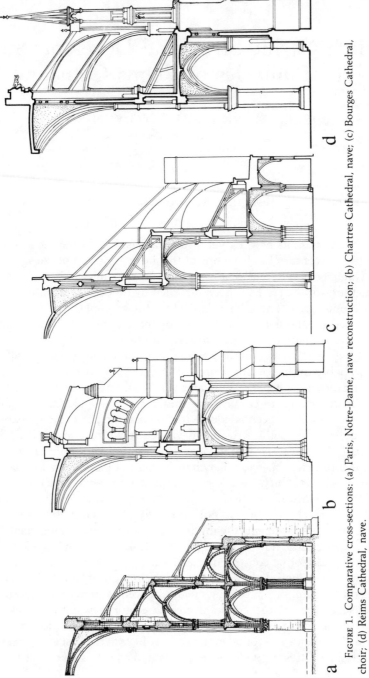

FIGURE 1. Comparative cross-sections: (a) Paris, Notre-Dame, nave reconstruction; (b) Chartres Cathedral, nave; (c) Bourges Cathedral, choir; (d) Reims Cathedral, nave.

126

tion to address the effects of wind loading. Structural modeling validated the structural feasibility of the new archaeological reconstruction, but it also revealed several regions of the buttressing system that might have demanded frequent maintenance. Recognition of these potential problem areas, along with the more generally accepted desire for increased interior light could well have contributed to the early thirteenth-century decision to undertake major alterations. Not surprisingly, this experience seems to have affected the design of the structure of the two giant buildings whose construction, begun in 1194–95, closely followed Notre-Dame, the cathedrals of Chartres and Bourges.

It had long been thought that the design of Chartres was established by the realization of the full potential of the flying buttress. From a modern technological viewpoint, however, the Chartres buttressing has been described as relatively ponderous, even clumsy; moreover, a tier of unsubstantial, largely ineffective upper flyers (FIG. 1b) was hastily erected during the last stages of construction.[3] The buttressing of Bourges, on the other hand, was shown to be simple, light and somewhat daring, but structurally sound.

Archaeological analysis of Bourges revealed that its original buttressing scheme was also altered during the course of construction.[4] The steep second tier of flyers supporting the high clerestory (FIG. 1c) apparently was an addition. We concluded that these efficient, highly sloped flying buttresses were brought into being as part of a pragmatic redesign of the cathedral structure in response to an increased awareness of wind effects on tall buildings. Such knowledge, before the full raising of Chartres or of Bourges itself, could have come only from Paris. Indeed, the added upper flyers of Chartres are now seen to have been erected in response to the problems at Notre-Dame. If not fully effective at Chartres, they appear nonetheless to have pointed the way for efficacious flying buttress placement in the mature High Gothic — a first, lower tier of flyers positioned to resist the outward thrust of the stone vaulting over the nave, and a second, upper tier of flyers bracing the high clerestory wall and the tall timber roof above the vaults against wind loadings — as exemplified in that next vast essay in Gothic architecture begun in 1210, the cathedral of Reims (FIG. 1d).

The classic phase of the High Gothic is generally agreed to have been established in the design of Reims by its first architect, Jean d'Orbais. As was typical at the time, construction of the cathedral started at the eastern end. In 1241, the choir and transept were first placed in service: and by the 1250s, five western bays, of the final total of eight, were standing.[5] Local political instability at Reims seems to have slowed the pace of construction so that three succeeding architects took charge of the project until it was finally com-

FIGURE 2. Reims Cathedral. View from north-west.

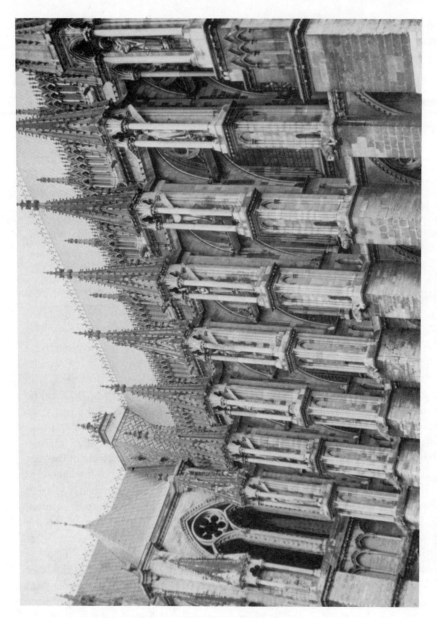

FIGURE 3. Reims Cathedral. Buttressing of the nave clerestory. (Photograph by R. Mark)

pleted in 1289; their work, however, held generally faithful to the Jean d'Orbais design, giving the entire building a remarkably unified aspect (FIG. 2).

THE QUESTION OF PINNACLE

A key element in the exterior visual massing of Reims is the array of great pinnacles that top the buttresses (FIG. 3). Pinnacles had been used earlier on buttresses, but never so emphatically (cf. FIG. 1b showing the residual pinnacle at the base of the upper flyer of Chartres with that of Reims, FIG. 1d). According to Jean Bony:

> These tall spirelets, each sheltering the winged figure of an angel, are not merely a lovely three-dimensional motif which gives firmness and unity to the upper part of the buttresses, nor is their end simply to accentuate the cadence of the bays: still more important is the horizontal continuity they create all around the building by their succession. . . . Reims is best understood from a distance . . . the design appears more than ever dominated by the motif of the pinnacled buttresses which, through their precision of form and their crystalline density, express within themselves the laws that organize the whole structure.[6]

It is doubtful that the influential, early twentieth-century architectural critic, Pol Abraham, would have taken exception to Bony's elegant description; but Abraham did cite Reims in a list of buildings whose pinnacle placement he considered structurally detrimental.[7] According to Abraham, the pinnacle, in addition to its beauty, might be a useful structural device only if it augmented the stability of a buttress—as determined by a vertical line taken through its center passing also through the center of gravity of the buttress. Accordingly, the placing of a pinnacle close to the exterior edge of the buttress would somewhat diminish stability. Of course he was correct in this, even though the long-time presence of the affected buildings attests to the fact that they are not lacking in stability; but Abraham did not take into account that the weight of a pinnacle can also provide additional compressive forces to the stones below it to further consolidate them and thereby help to prevent their lateral sliding, or shearing, under the action of thrust from the flying buttress. This function, moreover, does not depend on the exact position of the pinnacle on the pier buttress.

A similar observation that Abraham made about the pinnacles of the later cathedral of Amiens (begun 1220) led us several years ago to model that building's nave structure. On so doing, we discovered still another possible structural role for the externally placed pinnacle. When the Amiens model was tested under the action of simulated wind and dead-weight loads without the pinnacles, localized tensile forces appeared along the outer edge of the

FIGURE 4. Amiens Cathedral. Buttressing of the nave clerestory. (Photograph by R. Mark)

leeward buttress, just below the pinnacle base. Adding the pinnacle weight reversed these forces, canceling the tension.[8] We concluded from this that the pinnacle of Amiens acts as a prestressing element that prevents tension in the buttress. Furthermore, the sophisticated placement of the pinnacle on the Amiens buttress led to the realization that the medieval builder must have used observation of tensile cracking of the weak lime mortar to guide his design. We theorized that local cracking had been discovered following a storm of high intensity during the course of construction, before any of the pinnacles had been erected. The experience with modeling Amiens, however, could not be directly applied to Reims because of a number of differences between their structures. Amiens is significantly taller and more lightly constructed; its buttresses are solid, without aediculae sheltering statues of winged angels (FIG. 4); and the size of its pinnacles relative to the height of the buttresses is less than at Reims. Our expectation was that the Reims pinnacles also play a dual role of art and of structure, but there could be no certainty of this until Reims itself was modeled.[9]

STRUCTURAL MODELING

The rationale for small-scale photoelastic modeling of masonry buildings along with appropriate scaling laws is described in Mark (1982).[10] A brief summary of the basis of the engineering technique follows.

Analysis of the long vessels of Gothic churches is facilitated by their repeating, modular bay design. The buildings may be thought to be supported by a series of parallel, transverse frames consisting of the principal load-bearing structural elements: piers, buttresses, lateral walls and ribbed vaults. A second, important simplification is based on the assumption that the structural forces within the masonry "frame" are distributed as they would be in an equivalent frame constructed from an elastic, homogeneous material. This assumption has been shown to be adequate for predicting structural behavior in tests of reinforced concrete structures subjected to service loadings even though concrete is notoriously inelastic, compositionally inhomogeneous, and subject to tensile microcracking. For the simplification to be applied to masonry, it must also be assumed that the entire frame is undergoing compressive action: that is, that all of the individual stones are pressed against adjacent stones by interior forces. This assumption coincides with criteria for successful masonry performance because the tensile strength of medieval mortar is almost nonexistent; hence, structural continuity cannot be maintained if any substantial amount of tensile stress is present. Even small tensile stresses can cause cracking and begin a process of local disintegration, especially on the exterior of a masonry building subject to weathering. Our previous studies have indicated that compressive stresses prevail throughout Gothic buildings and that there are usually only a few highly localized regions of tension.

For photoelastic modeling, stress-free, epoxy models are loaded by arrays of weights representing the distributions of wind and the dead-weight forces acting on the prototype building. The model tests are performed in a con-trolled circulation oven where the epoxy is brought to a rubbery state (at about 140°C) and then slowly cooled, restoring the model to its room-temperature glassy state. Relatively large model deformations that took place at the higher temperature are "locked in" after cooling so that the loadings may be removed, with negligible effect. The unloaded model, now viewed through polarizing filters, displays interference patterns which, with calibration and scaling theory, can predict the force distributions in the full-scale structure. As general force distributions are sought rather than localized stress concentrations (for which the modeling method is often applied, for example in airframe analysis), no effort was made to *detail* the cross-sections of component structural elements; the action of the cross-ribbed vault, a three-dimensional structure, was simu-lated in the model as a planar arch. The heavily loaded foundations were also assumed to give complete fixity to the bases of the piers (*i.e.*, no deformation is permitted at ground level).

The structural system of the nave of Reims was modeled at 1:150-scale. The nave configuration was largely based on Dehio and Bezold,[11] with some adjustments from other sources. Dead-weight loadings and their distributions were computed from considering the volumes of stone in the elements com-prising a bay and taking the specific gravity of the stone as 2.4. The weight of a buttress pinnacle which consists in detail of a large, central, octagonal spire, slotted and hollowed out so that its net volume is about 70 percent of the gross, and four surrounding smaller spires of triangular section (FIG. 5) was estimated to be 29 tons (metric).[12] The weight of the statue, including its base, entablature and surrounding columns was estimated to be 23 tons. In performing these calculations, an interesting discovery was made: the weight of stone removed from a buttress to create the aedicula for housing the statue was estimated to be 54 tons, a figure very close to the combined weight of pinnacle and statue, 52 tons, and giving a net change in the total buttress weighting of practically zero.

Scaled dead-weight loadings were applied to the model in a similar pattern as the full-scale loadings in a ratio of 1:125,000 in two tests: the first without the simulated weights of the pinnacle and statue assemblies, and the second, with the full dead weight of all loadings. During these tests also, the model material was calibrated for optical activity with a controlled-stress specimen, and test data was photographically recorded.

In a third test, the model was loaded by simulated wind loads at 1:45,500-scale (the earlier patterns are "erased" when the model is reheated to test tem-perature). The prototype wind loading was based on long-term wind data for the Paris region from meteorological records, which indicated a possible max-

FIGURE 5. Reims Cathedral. Buttress pinnacles. (Photograph by R. Mark)

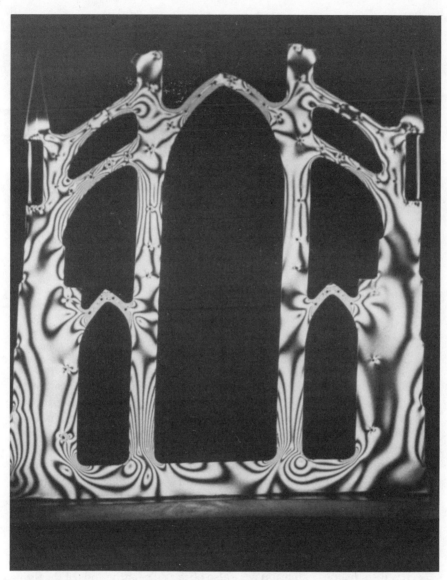

FIGURE 6. Photoelastic interference pattern in model of Reims Cathedral nave under simulated wind loading. (Photograph by Huang Yun-Sheng)

TABLE 1

Reims Cathedral: Buttress Stress

Location	Test Number	Loading Condition	Stress (kg/cm)
Windward buttress, within aedicula	(1)	dead weight, without pinnacle/statue	0.1 compressive
	(2)	full dead weight	0.6 compressive
	(3)	maximum wind	0.9 tensile
	(1) + (3)	dead weight (w/o pinnacle) + wind	0.8 tensile
	(2) + (3)	full dead weight + wind	0.3 tensile
Leeward buttress, below aedicula	(1)	dead weight, without pinnacle/statue	0.9 tensile
	(2)	full dead weight	0.6 compressive
	(3)	maximum wind	0.1 tensile
	(1) + (3)	dead weight (w/o pinnacle) + wind	1.0 tensile
	(2) + (3)	full dead weight + wind	0.5 compressive

imum wind speed at the 58 meter elevation of the Cathedral roof peak of 145 km/hr. The distribution of wind loading on the bay structure was estimated from wind-tunnel test data derived from a similar building form and using appropriate gust factors.[13] The interference pattern resulting from the third, wind load test is shown in FIGURE 6.

Combining the model results from the first and third tests (representing dead weight and wind, but with pinnacles and statues unweighted) and from the second and third tests (full dead weight and wind) and scaling these to the full-scale structure revealed, as in the previous tests of Gothic buildings, that compression prevails. Maximum compression stress occurs at the base of the piers: 17 kg/cm² under quiescent atmospheric conditions, and 23 kg/cm² when including the effect of high wind.

A few local regions of tension were also observed near the ends of the flying buttresses (typically found — particularly under the action of high winds) and along the external edges of the buttresses, within and just below the aedic-ulae. The flying buttress stresses were unaffected by adding the pinnacle and statue weights, but amelioration of the tensile stresses in the other regions was indicated when the weights of the pinnacles and statues were activated. Within the aedicula of the windward buttress under the action of high winds, the small tensile stress present without the weight of pinnacle and statue is reduced by 60%. Just below the aedicula of the leeward buttress, the tension which is caused mainly by dead weight and is actually little affected by wind action, is completely changed over to compression (TABLE 1).

All of the magnitudes of stress are moderate. The maximum compression at the base of the piers (23 kg/cm²) is comparable, for example, to a value of 34 kg/cm² for the piers of Amiens Cathedral under similar conditions of loading.[14] Wind is a more critical factor for the taller structure of Amiens and accounts for much of the additional compressive stress in its piers. Yet at Reims, high winds are the major source of tensile stresses caused by bending at the ends of the flying buttresses. These could reach values well above the usually taken, maximum threshold of mortar cracking (2 kg/cm²), a fact that corresponds to the necessary periodic maintenance of the flyers in almost all Gothic construction. The low magnitude of buttress tensile stresses (TABLE 1) indicates that these are not very critical even without the weight of the pinnacle and statue. Nevertheless, the builders might still have been made aware of their existence by seeing small fissures in the newly set mortar. The additional weights of the pinnacles and statues, according to the test results, would certainly have removed any trace of such cracking.

GOTHIC STRUCTURAL DESIGN

The detailed study of the nave of Reims afforded corroboration of certain earlier observations as well as some fresh insights concerning the approach to structure taken by the Gothic master builders. First, the role of the upper flying buttress in the classic High Gothic configuration was again confirmed: it acts mainly to resist high wind loadings on the church superstructure. Second, it appears that some care was taken to maintain a constant weighting on the exterior of the buttresses at Reims—by hollowing out the pinnacle so that its weight plus that of the statue assembly would equal the weight of material removed in forming the aedicula. This observation is substantiated by the presence of *solid* pinnacles in regions of the cathedral where they would play no structural role as, for example, along the west facade (FIG. 7). This balancing of the buttress pinnacle weight indicates that Reims' designer in some intuitive way shared Pol Abraham's anxiety over maintaining buttress stability. Indeed, Reims should be removed from Abraham's "endangered" list because, in effect, its pinnacled buttress maintains almost exactly the same stability as that of an equivalent solid buttress.

FIGURE 7. Reims Cathedral. "Decorative" pinnacles on the west facade. (Photograph by R. Mark)

In light of our analysis, the Reims designer's apparent anxiety for the stability of the buttresses points to his being overly cautious. Conservative structure is in fact a characteristic of Reims. While somewhat more refined than that of Chartres (cf. FIGS. 1b and 1d; for example, the heavy spur buttressing below the side aisle roof of Chartres is absent in Reims), it is still rather heavy compared with that of Bourges (cf. FIGS. 1c and 1d) or with the taller, more lithe structure of Amiens. The different treatment of the pinnacles of Amiens provides a further indication of the differing approaches to structure of their respective designers. The pinnacles of Amiens have been placed, without weight-balancing, along the exterior edge of the solid buttresses in a bold, yet evidently sound structural treatment. In both buildings, local tensile stresses in the buttresses are reduced or eliminated by the action of the pinnacles, but at Reims, subjected to lower wind loadings, these stresses would have been less crucial. This observation, as well as the elaborate sculptural treatment of the buttress, points to the conclusion that the pinnacle arrangement of Reims was predetermined, rather than being any kind of immediate solution to a pressing structural problem.

The buttress pinnacles of Reims do play a dual role of art and of structure, but the structural role is of far less moment than at Amiens. The placement of the pinnacles above the open aediculae and especially the effect of the laterally spreading wings of the statues below seem to express mainly another goal of that era: the visual dematerialization of Gothic structure.

ACKNOWLEDGMENTS

Support for this research has been provided by grants from the National Endowment for the Humanities and by the Andrew W. Mellon and the Alfred P. Sloan Foundations. We are indebted to these agencies for their support. We are also grateful to the architects Michel André and Bruno Chaffert-Yvart of the Service Départemental de l'Architecture in Reims for their kind assistance, and to Diane Rudenstine, a former graduate student in the School, Leon Barth, model maker, and Rod Rowland, Engineering Technician, for their help in modeling.

NOTES AND REFERENCES

1. R. Mark, *Experiments in Gothic Structure* (Cambridge, MA: MIT Press, 1982), pp. 55–56.
2. W.W. Clark and R. Mark, "The First Flying Buttresses: A New Reconstruction of the Nave of Notre-Dame de Paris," *Art Bulletin* LXVI(1) (1984): 47–65.
3. R. Mark, "The Structural Analysis of Gothic Cathedrals: Chartres vs. Bourges," *Scientific American* 227(5) (1972): 90–99.
4. R. Branner, *La Cathédrale de Bourges et sa place dans l'architecture gothique* (Paris/Bourges: Editions Tardy, 1962), pp. 52–53.
5. W.S. Stoddard, *Monastery and Cathedral in France* (Middletown, CT: Wesleyan University Press, 1966), p. 205.
6. J. Bony, *French Gothic Architecture of the 12th and 13th Centuries* (Berkeley, CA: University of California Press, 1983), pp. 271, 274.
7. P. Abraham, *Viollet-le-Duc et la Rationalisme Médiéval* (Paris: Vincent, Fréal & Cie, 1934), pp. 88ff.
8. Mark,[1] pp. 52–55.
9. Reims Cathedral suffered extensive damage during the First World War. Possibly some clues about the structural action of the elements of its buttressing system came to light during the course of extensive rebuilding, but if so, they do not seem to have been recorded.
10. Mark,[1] pp. 18–33.
11. G.G. Dehio and G. von Bezold, *Die Kirchliche des Abendlandes* (Stüttgart: Hildesheim, 1892–1901).

12. The Reims pinnacle aggregation seems to derive from similar patterns of full spires as found, for example, atop the western towers of the nearby Abbey church of Saint-Remi.

13. The total maximum wind loading (pressure and suction, and including gust factors) acting on a nave bay of Reims was estimated to be 105,000 kg, or six percent of the estimated bay weight, $1.7 (10)^6$ kg.

14. Mark,[1] p. 55.

The Function of Mechanical Devices in Medieval Islamic Society

GEORGE SALIBA

Department of Middle East Languages and Cultures
Columbia University
New York, New York 10027

MOST WRITERS dealing with mechanical devices in medieval Islamic techni- cal treatises have, in one way or another, raised the question of the utility of these devices, and have mostly concluded that they were more toys than useful machines. The terms used to describe these devices: "tricks," "automata," "amusing automata," and the like, are usually applied to the machines described by the most famous medieval Muslim engineer Badī' al-Zamān Abū al-'Izz Ismā'īl Ibn al-Razzāz al-Jazarī (fl. 1200).[1] His treatise has been known for some six decades now, and extensively studied by art historians who saw a fertile field of study in the illustrations of the machines that accompanied the text of Jazarī. The comparatively less famous book of Banū Mūsā[2] (the three brothers who became famous for their engineering works and their patronage of other scientists during the ninth century in Baghdad and Samarra), with its description of water fountains and dredging devices, but mainly drinking vessels and self-trimming lamps and the like, has also contributed to the general impression that the main function of these machines was for entertainment and amusement.

Other treatises dealing with monumental clocks, such as the work of Ridwān Ibn al-Sā'ātī (fl. 1200)[3] and the several works dealing with astronomical instru- ments such as equatoria and the like are usually studied in a different context and need not concern us here, for they are rarely confused with the books of Jazarī and Banū Mūsā that are under consideration now.

In this paper, the works of Jazarī and Banū Mūsā will be re-examined in an attempt to evaluate critically the validity of the conclusion mentioned above. Moreover, the purpose of such writings will be investigated against the back- ground of the general perception in Medieval Islamic times of the role and function of the mechanical sciences in general, and thus the role of such writings as those of Jazarī and Banū Mūsā. Finally, the social and political function of the toys itself will be examined, and new material will be brought to light to support the hypothesis of an uninterrupted production of marvelous machines and toys from Hellenistic times, and continuing well into Islamic times, but not recorded by either Banū Mūsā or Jazarī.

In an attempt to determine the reasons for which Jazarī wrote his book, two remarks should be made at the outset. First, the title of the book which Jazarī called *al-Jāmiʿ bayn al-ʿilm wa-l-ʿamal al-nāfiʿ fī ṣināʿat al-ḥiyal* (*A Compendium of Theory and Useful Practice in the Mechanical Arts*) has not been subjected to any detailed study, as far as I know. The subject matter of the book, as conceived by the author himself, has something to do with *Ḥiyal*, which should be understood in the sense of machine as related to Greek μηχᾰνή "a contrivance," a "device." To understand the full significance of the word, we should recall the use of μηχᾰνή in the Aristotelian *Mechanical Problems*,[4] where it is discussed in the following context.

The opening sentence of the *Mechanical Problems* is: "Remarkable things occur in accordance with nature, the cause of which is unknown, and others occur contrary to nature, which are produced by skill [τέχνη = Arabic *ṣināʿa*] for the benefit of mankind." As an example of things that seem to occur "contrary" to nature, the Aristotelian text later uses the lever device where the text notes: "It is strange that a great weight can be moved by a small force . . . ." "When, then, we have to produce an effect contrary to nature," the text continues, "we are at a loss, because of the difficulty, and require skill (τέχνη). Therefore we call that part of skill which assists such difficulties, a device (μηχᾰνή). Of this kind are those in which the less master the greater, and things possessing little weight move heavy weights, and all similar devices which we term mechanical (μηχανικά) problems." The text goes on to say that "these are not altogether identical with physical problems, nor are they entirely separate from them, but they have a share in both mathematical and physical speculations, for the method is demonstrated by mathematics, but the practical application belongs to physics."

A machine (Ḥīla) is therefore any device that allows one to overcome the natural resistance, and thus perform actions contrary to the natural tendency. In that sense of tricking, *i.e.*, tricking nature, the words machine, and Ḥīla (pl. Ḥiyal) are a translation of the Greek μηχᾰνή. The last sentence of the above quotation however, brings us to the second remark that should be made about the title of Jazarī's book. The intermediate position of the mechanical problems, between the mathematical sciences and the practical application, explains very well the term *Jāmiʿ* = "that which combines," "a compendium," as that which combines theory (*ʿilm*) and practice (*ʿamal*). What Jazarī seems to have thought he was doing was writing a book in the tradition of the Aristotelian *Mechanical Problems*. And, therefore, that book combines theory and practice in matters pertaining to mechanical arts.

Pursuing this line of Aristotelian thinking, Jazarī stated in his rationale for writing his book that his patron had asked him to commit to writing a description of the things he had made. The terminology that he used, however, is transparently Aristotelian, for he had his patron say to him: "*laqad ṣanaʿta ashkāl[an] ʿadīmat al-mithl wa-akhrajtahā min al-quwwat ilā al-fiʿl*" (You have made peerless models and brought them forth (*akhrajtahā*) from potentiality (*al-quwwat, δύναμις*) into actuality (al-fiʿl, ἐνέργεια).

Therefore, Jazarī must have thought of his work as a demonstration (*tamthīl*) of how things could be brought forth from potentiality into actuality by means of devices called *Ḥiyal* (μηχάνή). And since the models that he described were only samples of specific applications,[5] his book therefore combines both the theoretical principles and the practical applications.

Turning to the second important mechanical work that has survived from Medieval Islamic times, *i.e.* the work of Banū Mūsā, we find that we can not pursue this line of argument very far, for the surviving text has the innocent title *Kitāb al-Ḥiyal* (The Book of Mechanical Devices), with *Ḥiyal* (sg. *ḥīla*) to be understood as a good translation of Greek μηχάνή mentioned above, and lacks the introductory rationale that would have revealed the intentions of the authors behind the composition of their book. Instead, the book begins immediately by describing the construction of a drinking vessel known from the works of Heron of Alexandria (first century A.D.) and Philo of Byzantium (first century A.D.?) (more about this relationship with the Greek texts below). But a curious note on the margin of the page on which model 48 is drawn, puts the question of the function of these devices back along the same lines developed above. The note reads: "*taʿammal hādhihi al-ālah fīhā sabab[un] yamnaʿ min khurūjihā ilā al-fiʿl*" (Consider this instrument for it has a factor that does not allow it to come forth (*khurūjihā*) into actuality). Although this note may not have been written by the authors themselves, it still reveals the way their work was understood by their readers. And this marginal commentator in specific, must have been an engineer, like the authors, who either figured out the technical details of model 48, or tried to build it himself and failed, thus the remark that it could not be actualized. In either case, he also thought that he was bringing things into actuality.

This line of thinking is also Aristotelian. For it was Aristotle who defined the function of all art with the following statement: "All art (τέχνη) is concerned with coming into being, *i.e.* with contriving (τεχνάξειν) and considering how something may come into being which is capable of either being or not being, and whose origin is in the maker and not in the thing made."[6] The note that the commentator to the Banū Mūsā text used, is cast in the same terminology of "coming forth into actuality" as that used by Aristotle and Jazarī.

Further research concerning the use of such terminology proved to be very productive, for after some investigation it was found that the works of other engineers, which we shall refer to below, also employ exactly the same concepts.

In a badly preserved treatise of the eleventh-century Andalusian engineer, Ibn al-Murādī, dealing with mechanical devices and including mechanical clocks, the author refers to his work as dealing with "philosophical models" (ashkāl fāylasūfiyya) that "I brought forth [akhrā]jtuhā from nihility ('adam) into existence."[7]

Another engineer, Mu'ayyad al-Dīn al-'Urḍī (d. 1266), writing about 1250, spoke of the Indian circle as a practical method for determining the local meridian, as being "the best in matters that one wishes to bring forth (ikhrāj) from potentiality into actuality."[8] And in another treatise describing the astronomical instruments that the same engineer 'Urḍī had built at the Maraghah Observatory in North West Iran (during the years 1259–1261), he says: "As for the instruments that we invented and completed, we brought forth (akhrajnā) some of them into actuality (ilā al-fi'l) in full form (kāmila) and are to be found at the blessed observatory, while others we only made models (mithāl) for them."[9]

The last quotation contains not only the general perception of these mechanical devices as embodiments of principles existing only potentially that could be brought forth into actuality, but also gives us an idea of the method by which at least this engineer operated. Apparently, he would first make a model (mithāl) of the instrument or the machine, not unlike scaled-down modern day working models, and only after showing that it could be actually constructed, he would go ahead and build it in full form (kāmila). If this procedure is true, it would also explain why Jazarī was told by his patron that he wanted him to write a book in which he would describe what he alone made models of (tamthīl). It may also explain why some dimensions given in connection with some machines are not realistic: they could have resulted from mistakes in scaling from the model to real life or vice versa. Since we only have 'Urḍī's statement that such models did indeed exist, we must refrain from speculating about this procedure lest we be accused of reading too much into one sentence. The lack of archeological remains does not help the situation either, for had there been any remaining models of scaled-down instruments one would have read 'Urḍī's statement with greater confidence.

To go back and consider the place of these texts on mechanical devices within the general framework of the body of literature produced during medieval Islamic times, it would be useful to recall the statements of people other than the engineers themselves. The historian Ibn Khaldun (d. 1382), for example, refers to carpentry by saying that "it needs a good deal of geometry . . . in order to bring the forms (of things) from potentiality into actuality in the proper manner."[10] Hence he confirms the Aristotelian understanding of the place of a skill (τέχνη) such as carpentry, within the scale of knowledge and coming-to-be.

But the more significant statement about mechanical devices is that of the tenth century philosopher Fārābī (d. 950), who devoted a full chapter to the science of Ḥiyal (Mechanical Devices) in his Iḥṣā' al-'Ulūm (Enumeration of the Sciences). Since I do not know of a complete English translation of this chapter[11] and since it is of some interest for the status of other sciences in tenth-century Islam, I thought it would be worth translating it in full at this point. Fārābī says:

> The science of mechanics (Ḥiyal) is the knowledge of the procedure (wajh al-tadbīr) by which one applies (muṭābaqa) all that was proven to exist in the mathematical sciences (ta'ālim) that were mentioned above in statements (qawl) and proofs unto the natural bodies, and (the act of) locating (all that), and establishing it in actuality (bi-l-fi'l). The reason for that is that these (mathematical) sciences concern themselves with lines, surfaces, volumes, numbers, and all their subject matter is intelligible on its own in isolation from the natural bodies. When one wants to locate (these ideas that form the subject matter of the mathematical sciences) and willfully exhibit them (by means) of a craft (ṣun'a = τέχνη) in the natural bodies that are perceptible to the senses, one needs a force (quwwa) through which he proceeds to establish them in (these bodies) and to apply (these ideas) to (these bodies). For the material and perceptible bodies have special conditions that prohibit them from accepting (the ideas) that were demonstrated by proofs from being located in them as one pleases to do. On the contrary, these natural bodies have to be prepared to accept what one seeks to establish in them, and one has to contrive to remove the obstructions (an yatalaṭṭaf fī izālat al-'awā'iq).
>
> The sciences of mechanics are therefore those that supply the knowledge of the methods and the procedures by which one can contrive to find this applicability and to demonstrate it in actuality (bi-l-fi'l) in the natural bodies that are perceptible to the senses.
>
> Of these mechanical sciences are the many arithmetical ones including the science known to the people of our times as the science of Algebra (al-jabr wa-l-muqābala), for it partakes of arithmetic and geometry. It also includes the procedures by which one brings forth (istikhrāj) the numbers which ought to be dealt with in accordance with the principles laid down by Euclid in Book X of the Elements, as to their being rational (muntaqa = expressible) or surds (ṣumm = inexpressible), or in accordance with others that were not mentioned by him in that book. For since the ratio of the rational and the surds to one another is like the ratio of one number to another, then each number is the coun-

terpart (*naẓīr*) of some magnitude, be it rational or surd. If one could bring forth these numbers that are the counterparts of the ratios of these magnitudes, then these magnitudes would have been brought forth in one way or another. For that reason, some numbers are taken to be rational as the counterparts of the rational magnitudes while the others that are surds are counterparts of the irrational magnitudes.

Among them (*i.e.* the mechanical sciences) also, are the many geometric (or engineering, *handasiyya*) mechanical sciences, such as:

The art of overseeing constructions (*ṣināʿat riʾāsat al-bināʾ*).

The devices for determining the areas of bodies.

The devices used in the production of astronomical and musical instruments, and in the preparation of instruments for many practical crafts (*ṣanāʾiʿ*) such as bows and arrows and various weapons.

The optical devices used in the production of instruments that direct the sight in order to discern the reality of the distant objects, and in the production of mirrors upon which one determines the points that reverse the rays by deflecting them (*taʿṭufahā*) or reflection (*taʿkusahā*) or refraction (*taksurahā*). With this, one can also determine the points that reverse the sun's rays unto other bodies, thus producing the burning mirrors and the devices connected with them.

The devices used in the production of marvelous objects, and the instruments for the several crafts (*ṣanāʾiʿ*).

These and their likes are the mechanical sciences which (in turn) are the principles (*mabādiʾ*) of the civil and practical crafts that are applicable to bodies, shapes, positions, order, and assessments such as in the crafts of masonary, carpentry and others.

These then are the mathematical sciences (*taʿālīm*) and their divisions.[12]

Of the many interesting ideas expressed in this passage such as the explicit statement concerning the novelty of the field of algebra as having no counterpart in the classical tradition, and the implied statement that surds could be treated as numbers (again a novel concept), the most important one for our immediate concern is the line of argument that sheds light on the perception of the role of the mechanical sciences in medieval Islamic times. These mechanical sciences are unequivocally defined as the means by which one could bring forth ideas or principles that exist in potentiality (δύναμις) into the observable actuality (ἐνέργεια). The inclusion of the science of Algebra among the mechanical sciences is very revealing, especially in the context discussed by Fārābī. The reason he thinks that Algebra should be included under this category is that Algebraic manipulations allow one to bring forth the quality of number from the surds (δυνάμεις), where it exists only in potentiality, into actuality, so that they become expressible (*munṭaq*). This interpretation of the mechanical sciences agrees very well with the usage employed by the writers on mechanical sciences themselves.

To summarize therefore, it seems clear that the perception of the mechanical sciences (*Ḥiyal*) in medieval Islamic times derived ultimately from the Aristotelian definition of skill (τέχνη). The subject matter of these sciences was supposed to be the actualization of principles that existed in potentiality. Hence any of these devices, described in the mechanical sciences, is in theory an actualization of some principle or other.

But could this general theoretical framework explain all the devices described by Jazarī or Banū Mūsā? Are all of these devices amusing applications of principles — whether they are to be interpreted as toys or not — or are they useful machines and instruments seen as satisfying social and technological needs? To answer either one of these questions, one has to draw some distinction between the frivolous toy — that may be technologically very sophisticated in the sense of being an actualization of a theoretical principle or a combination of principles in the sense discussed above — and the useful machine which need not be anything more than a simple milling wheel driven by the force of an animal or that of a river. But this distinction between the toy and the useful machine should never conceal the fact that a free-market demand for either could give rise to a higher production irrespective of the utility concerned — if utility is to be understood as generating further production. There is obviously a distinction between a technological tradition that leads to the production of a steam engine, and that which leads to the production of a gadget that operates by the power of steam.

From that perspective, medieval Islamic writings on mechanical devices describe both the toy and the useful machine. Both Banū Mūsā and Jazarī include in their respective texts detailed descriptions of drinking pitchers that seem to have no other purpose except entertainment; these may be understood as toys or marvelous devices. But both texts include a variety of fountains that change the shape of their water-jet, and could be seen as very useful devices. Combination locks, water clocks, measuring vessels, water-lifting devices, and the like seem to be directed at real-life needs.

To concentrate on the "toys" for a while, it may be worth noting that in these toys some real technological advances could also be made. Take, for example, the constant-level water basins described by Philo of Byzantium,[13] which operate by the siphon principle. Compare that with model 75 as described by Banū Mūsā,[14] which performs the same function, namely, to keep the water in a basin at a constant level. Banū Mūsā's device, however, is operated by a float and a crank rather than by the siphon action. Technologically, the crank, once introduced, becomes a very productive element for it gets used in several other machines, starting with Banū Mūsā themselves.[15] One need not stress any further the useful functioning of such a device in later constructions of "useful" machines.

Consider, on the other hand, the purely entertaining devices such as the tree of the chirping birds described by Heron of Alexandria.[16] Ingenious as it is, no such device is described by Banū Mūsā, or by Jazarī in spite of the fact that Banū Mūsā take their inspiration from Heron's work for some twenty devices that they describe, especially the drinking pitchers and the self-trimming lamps. This does not mean that the construction of such devices was beyond the technical ability of medieval Muslim artisans. In fact we have a tenth-century report of the presence of such a tree of chirping birds at the caliphal palace in Baghdad.[17] A similar tree was also found near the throne of an eleventh-century dynast in Central Asia.[18] From another source, we also know that Banū Mūsā themselves maintained their political influence at the court of the caliph al-Mutawakkil (847–861) because of the latter's fascination with automata that Banū Mūsā made for him.[19] The same source, however, also reports their real-life engineering enterprises in irrigation that they were also supposed to perform for the court. Yet, neither the text of Banū Mūsā, nor that of Jazarī, contain any toys of the type described by Heron and which must have existed at the medieval Islamic courts.

The written tradition on mechanical devices in medieval Islamic times was therefore seen as an expression of a science whose subject matter is the actualization of principles that exist in potentiality. At times little distinction was made between this actualization as embodied in a "toy" and an actualization that is embodied in a "useful" machine. But at other times, the written tradition did not see it fit to describe purely entertaining toys, although these toys were continuously produced by medieval Muslim artisans.

The main thrust in the two most famous texts from that tradition is to describe models of fountains, clocks, water-lifting devices, combination locks, and architectural edifices such as iron gates and monumental water clocks, all intended to describe what was actually constructed or suggest improvements over real-life machines. Only a minor part of these texts could be thought of as describing "toys."

It is only accidental that the illustrations of these texts, being artistically beautiful indeed, were first noticed by modern scholars as they surfaced in the art-collecting market before the appearance of the texts themselves. As a result, the main purpose of these texts was misunderstood, in spite of the contents of the texts and their expressed intentions.

But it is not accidental that functional machines were also intended by medieval engineers to be beautiful at the same time. The beautiful illustrations in the texts did indeed refer to machines and devices whose beauty was stressed by the engineers themselves. This tendency to combine functionality and beauty is further illustrated by any of the several hundred existing astrolabes from medieval Islamic times, where the precision and functionality of the astrolabe is never compromised by its beauty: most of these astrolabes are astonishingly beautiful pieces of art work. I would like to see the fountains with varying water-jets as another expression of this tendency.

NOTES AND REFERENCES

1. See, for example, R.J. Forbes, in *History of Technology*, Singer *et al.*, eds. (Oxford: Oxford University Press, 1956), p. 614, where he refers to Jazari's machines as "amusing automata;" D. Price, in *Technology and Culture* **16**:81(1975), calls them "philosophical toys;" A. Keller, in *Technology and Culture* **21**:232(1980), refers to Banū Mūsā's machines as "conjuring tricks, amusements to deceive;" and most recently E. Atil in *Renaissance of Islam: Art of the Mamluks*, (Washington, D.C.: Smithsonian Institution Press, 1981) p. 256, where she refers to Jazari's machines as "princely toys."
    The text of Jazari itself was first translated by D.R. Hill as *The Book of Knowledge of Ingenious Mechanical Devices* (Dodrecht/Boston: Reidel, 1974) and edited by A.Y. Hasan as *Al-Jāmi' bayn al-'Ilm wa-l-'Amal al-Nāfi' fī Ṣinā'at al-Ḥiyal* (Aleppo: Institute for the History of Arabic Science, 1979).
2. *Kitāb al-Ḥiyal*, A.Y. Hasan, ed. (Aleppo: Institute for the History of Arabic Science, 1981), which has been translated by D.R. Hill, as *The Book of Ingenious Devices* (Dodrecht/Boston: Reidel, 1979).
3. For a recent study and a partial English translation of Ibn al-Sā'ātī's work, see D.R. Hill, *Arabic Water-Clocks* (Aleppo: Institute for the History of Arabic Science, 1981) pp. 69–88.
4. Aristotle, *Mechanical Problems*, trans. W.S. Hett, Loeb edit. 1936, pp. 331f.
5. At the end of category IV of Jazari's text, the author adds the following remark:
    I say that this kind of water-fountain and whistling device – which is in ten models (*ashkāl*) – can all be produced by one marvelous and amazing device with which one can produce several other models of various forms. I did not mention any of them on account of their multiplicity. But I will explain the principle (*aṣl*) of this device in order that any one, having the slightest concern with this science (*'ilm*) and craft (*'amal*), can derive from it several works (*a'māl*). I will illustrate that with one clear picture from which one can derive these works.
    Jazari, *Jāmi',*[1] p. 436.
    It is interesting to note that this section of Jazari's text is missing from the Bodlean Manuscript, Graves 27, which was used as the basis for Hill's translation, thus leaving the accompanying illustration, which has survived in that manuscript, inexplicable. Now that the text is edited, one can understand the functioning of that device.
    Jazari's statement, however, illustrates clearly his understanding of his text as a series of specific applications of such principles that could be multiplied by any one who knows both the theory and the practical application of these principles.
6. *Nicomachean Ethics*, VI, 4, 1140a,10, trans. W.D. Ross.
7. For the identification of the name of this author and the reconstruction of the word *akhrajtuhā*, see A.I. Sabra, "A Note on Codex Bibliotheca Medica-Lorenziana Or. 152," *Journal for the History of Arabic Science* **1**:276–283(1977), esp. p. 279.
8. *Kitāb al-Hay'ah*, Bodlean Manuscript, Marsh 621, fol. 200r.
9. S. Tekeli, "Al-Urdi'nin 'Risalet-ün fi Keyfiyet-il-Ersad' Adli Makalesi," *Arastirma* **VIII**(1970):135.

10. Ibn Khaldun, *Al-Muqaddimah*, trans. F. Rosenthal, Bollingen Series XLIII (Princeton, NJ: Princeton University Press, 1958) vol. 2, p. 365.

11. There is a partial English translation of this chapter by Marshall Clagett and Edward Grant based on the medieval Latin translation of Fārābī's *Iḥṣā'* by Gundisalvo, which was recently published in *A Source Book in Medieval Science*, E. Grant, ed. (Cambridge, MA: Harvard University Press, 1974), p. 76.

The Arabic text of Fārābī's *Iḥṣā' al-'Ulūm* was edited by O. Amin (Cairo: al-maktabah al-Anglo-Misriyyah, 1968). The chapter translated here is from that edition.

12. *Iḥṣā'*,[11] pp. 108–110.

13. See A.G. Drachmann, *Ktesibios, Philon and Heron* (Copenhagen: E. Munksgaard, 1948) pp. 50f.

14. Hill,[2] (Banū Mūsā) trans. p. 192; Hasan,[2] p. 281.

15. *Cf.*, for example, Banū Mūsā, Hill[2] and Hasan,[2] Models 78, 79, 80, 81, 82, 83, 85.

16. *The Pneumatics of Hero of Alexandria: A Facsimile of the 1851 Woodcroft Edition*, introduced by M.B. Hall (London/New York: Macdonald and American Elsevier, 1971), model 15, p. 31.

17. Ibn al-Jawzī's, *Muntaẓam* (edition Hyderabad, 1938–1939, vol. 6, p. 144) has the following account:

In the year 305 H. (= A.D. 917), there came a messenger from Byzantium seeking a treaty and an exchange of prisoners. The messenger was a young man accompanied by an older man and some twenty servants. They were received in posh surroundings for a few days until they were brought to the caliph's palace to which he passed by the review of the armed forces of 160,000 knights and infantrymen, followed by several kinds of slaves and servants. The messenger almost descended by the house of the doorkeeper which had already impressed him to think that it was the caliph's house. He was advised that it was only the doorkeeper's house. Then he was brought to the Vezir's house which he had not doubted that this time it was the caliph's house. He was told that it was only the Vezir's. When he finally arrived at the caliph's palace, he was taken around the palace to watch the festive decorations, the domesticated wild animals eating from peoples hands, and some hundred lions or so with their trainers. *Then he was brought into the house of the tree which was placed in the midst of a pool that had clear water, and which had eighteen branches and each branch had several branchlets adorned with birds of every kind made of silver and gold, and most of the branches were made of silver while others were of gold. [The branches] were swaying with their multicolored leaves, and each kind of these birds was chirping.* [emphasis added]. Only after that, he was taken into the glorious room (lit. paradise) which was also heavily decorated with gold silk, where he finally met the caliph who was attended by his translator and Vezir.

18. *Muntaẓam*,[17] vol. 9, pp. 110, reports the following account:

During the year 492 H (= A.D. 1098) Kiyā, al-Ṭabarī, the jurist, was sent to the Central Asian dynast Ibrāhīm b. Mas'ūd b. Maḥmūd b. Subukhtakīn where he saw the sultan surrounded by gold and silver statues. After that, he was taken around the palace where he saw a throne surrounded by birds who would flap their wings when the king sat on it. The messenger, being a jurist, chastised this king for his pomp.

The same source, *Muntaẓam*, vol. 10, p. 148, also reports that during the time of celebrations several automata were usually set up in public places, each bazaar group of craftsmen exhibiting their own automata. This kind of festivities called *ta'līq*, literally "mounting up," was celebrated when a monarch assumed power or when an heir was announced, or the like.

From a later source, Ibn Ḥajar al-'Asqalānī (A.D. 1449) *Inbā' al-Ghamr bi-abnā' al-'Umr*, 9 vols., edition Hyderabad, 1967–76, vol. 1, p. 156) reports the following:

The gifts of the Byzantine Ruler of Istanbul arrived that year (777 H = A.D. 1375) [to Cairo]. Among them was a box that included some moving figurines. At the end of each hour of the night, these figurines would play some music. And at the passage of each degree [of the equator] a pellet would fall.

19. Ibn Abī Usaybi'ah, *Ṭabaqāt al-Aṭibbā'*, A. Muller, ed. (Konigsberg, 1884, vol. I, p. 207).

Fevers, Poisons, and Apostemes: Authority and Experience in Montpellier Plague Treatises

MELISSA P. CHASE
Department of the History of Science
Harvard University
Cambridge, Massachusetts 02138

In 1348, Europe was ravaged by the Black Death, the first in a series of outbreaks of plague that would continue for the next three centuries. Medieval physicians, confronted with these devastating epidemics, responded in part by writing hundreds of treatises devoted to the pestilential outbreaks.[1] Most of these treatises were practical in nature, being little more than collections of remedies to prevent and cure the pestilence. However, many of them, implicitly or explicitly, attempted to explain the epidemics within the theoretical framework of medieval medicine.

Medieval understanding of disease was dominated by the Hippocratic-Galenic tradition, which emphasized humoral and qualitative balance. In this tradition, a healthy body was seen as being in a state of equilibrium between the primary qualities or humors, while disease resulted from upsetting that balance. In other words, there was fundamentally one disease — imbalance. To be sure, if one peruses medieval medical texts, one finds a rich taxonomy of disease. Medieval physicians classified *diseases* on the basis of symptoms, causes, and the parts of the body affected. Collections of symptoms were given names and were written about as though they were distinct entities. Despite these taxonomic efforts, when medieval physicians theorized about health and disease, humoral balance and imbalance were the paramount explanatory devices.[2]

The plague outbreaks presented a challenge to late medieval medicine. Medieval physicians perceived the plague epidemics as a novel phenomenon, one not experienced by their Greek and Arabic authorities. Several late fourteenth-century physicians associated with the medical faculty at Montpellier made this perception explicit. Johannes Jacobi wrote that since the pestilence invades us more frequently than the ancients, few of the ancient authorities had experienced the disease.[3] Guy de Chauliac believed that the ancients had had experience with epidemics, but that those earlier outbreaks had been local,

153

unlike the universal scourge of his day.[4] Raymundus Chamelli de Vivario asserted that the ancients had not fully understood the causes of the epidemics, and thus could neither fully comprehend the disease, nor properly cure it.[5]

The plague literature and its importance for medieval medicine have not escaped the notice of historians, but previous studies have failed in one important respect: they limit their sources to the works written at the time of the Black Death.[6] But if one is interested in the process by which medieval physicians assimilated observed phenomena to medieval medical theory, one must examine treatises written over a longer period of time in order to determine the effects of growing familiarity with the disease.

In this essay I will examine plague treatises written between the time of the Black Death and the mid-fifteenth century by physicians associated with the medical faculty at Montpellier. Approximately 25 separate treatises or parts of larger works of Montpellier provenance on the plague epidemics have come down to us. This literature can roughly be divided into three chronological groups, clustering around periods of plague outbreaks. The first collection of works was written at the time of the Black Death, the second between 1361 and 1381, and the third during the first two decades of the fifteenth century.

The Montpellier plague literature is a particularly significant subject for analysis for two reasons. First, the medical faculty was a prominent one during the late Middle Ages; although it was comparatively small, students were attracted to Montpellier from all over Europe, and many Montpellier masters were engaged by the papal court, kings and nobles.[7] In addition, since all the authors appear to have practiced in the Midi and Catalonia, their experiences with the epidemics were probably comparable.[8]

The first group of Montpellier physicians to write on the plague is quite obscure. The longest work was written by an anonymous practitioner, apparently in response to the celebrated consultation of the Paris medical faculty.[9] Another treatise was prepared by Alphonse of Cordova, who describes himself as a master of arts and medicine.[10] The most celebrated of these early authors was Bernard Alberti, who appears to have had the epidemics in mind during his discussion of pestilential fevers in his compendium on fevers, the *Practica provectis in theorica supra prima fen quatuor Canonis*.[11]

The first generation of Montpellier plague authors considered the epidemics to be outbreaks of pestilential fever, whether or not they called it such. In medieval medical theory, fever was not merely a symptom, but a category of disease characterized by excess heat within the body.[12] Fevers were subdivided into three main types distinguished according to the part of the body first afflicted: hectic fever first arose among the solid members, ephemeral fever among the spirits, and putrid fever among the humors. Pestilential fever occupied an anomalous position within this classification, for what distinguished it from other types of fever was not where it arose but what caused it.[13] Under the influence of the heavenly bodies, the ambient air became putrefied through the admixture of "bad vapors" (*vapores mali*); this putrefied air then entered the body, traveled to the heart, and corrupted the complexion of this principal member, which, as in other fevers, generated excess heat, which is harmful to the body.[14]

The most elaborate description of the causes and clinical features of the pestilence is contained in the work written by the anonymous practitioner. Expanding upon the accounts of pestilential fever found in Avicenna and early fourteenth-century Montpellier authors, he maintained that the conjunction of Mars, Jupiter, and Saturn in Aquarius in 1345 drew vapors from the earth and water, which corrupted the surrounding air.[15] Once the putrefied air entered the body, it traveled through the veins to the lung, causing its motion to cease, which in turn heated the heart, thus producing a pestilential fever.[16]

The preceding description is little more than an elaboration upon the traditional account of pestilential fever found in Avicenna and the pre-plague Montpellier authors, albeit with a greater emphasis upon the celestial causes. On top of this traditional foundation, however, the Montpellier practitioner and his fellow authors constructed some novel features. First of all, without question the epidemic was thought to be contagious.[17] This represented a departure from earlier discussions of pestilence: Avicenna did not incorporate contagion as a feature of his discussion of pestilential fever, and although pestilential fever was usually counted among the infectious diseases by medieval physicians, this view did not appear in pre-plague discussions of pestilential fever at Montpellier.[18]

Another departure from previous discussions of pestilential fever was an emphasis upon apostemes, or swellings. Avicenna, and the Montpellier writers on fevers, had discussed the relation between apostemes and fevers in general, but as with contagion this had not been a feature of earlier accounts of pestilential fever.[19] In the wake of the Black Death, predominately an epidemic of bubonic plague, these Montpellier practitioners described the lesions that they variously called *bubones, apostema,* or *carbunculi.*[20] The appearance of these swellings was explained by humoral pathology: if there was a superfluity of humors in the liver, apostemes would form in the groin, if the superfluity was in the heart the swellings would appear under the armpits, and if the excess was in the brain they would appear behind the ears.[21]

The final novel feature found in the Montpellier literature on the Black Death is the role of poison. In the earlier discussions, pestilential fever resulted from corruption and putrefaction of the surrounding air through the admixture of noxious vapors arising from such foul places as swamps and latrines. Within the body, these putrid vapors caused the humors and other moistures (*humiditates*) within the body to putrefy. The authors of the plague treatises invoked this process with a subtle difference: the *humiditates* were not simply putrid, but poisonous. The anonymous practitioner used this concept of poisonous vapors to provide a mechanism for contagion: the poisonous fumes within the body rose to the brain and poisoned the visual spirit, which then passed out through the eye of the infected person to afflict someone standing nearby.[22] The author tried to persuade his readers of the reasonableness of this mechanism by analogy with the well-known legend of the basilisk, a mythical creature that kills those who look upon it by poisoning them through the eye, and the story of how Aristotle uncovered the plot of the maiden who was nourished on poison so that she might poison Alexander the Great through intimate contact.[23]

Alphonse of Cordova presented his own idiosyncratic discussion of the role of poison in the contemporary epidemics. He suggested that the epidemic had lasted too long to have been caused through natural means, that is, the impression of the heavenly bodies upon the primary elements, and thus must have been caused through malicious artifice.[24]

After the initial spate of literature around the time of the Black Death, there was a lull until the late fourteenth-century outbreaks of 1361, 1374, 1375, and 1381 which renewed interest in the subject. Some of the outstanding figures of the late fourteenth-century Montpellier medical faculty contributed to this plague literature, along with lesser known practitioners. In 1363, Guy de Chauliac discussed the epidemics of 1348 and 1361 in the section of his *Magna chiurgia* devoted to apostemes.[25] In the 1370s, Johannes de Tornamira and Johannes Jacobi, who vied with each other for the chancellorship of the medical faculty, and served as physicians to various members of the Avignon court, both wrote plague treatises and discussed the epidemics in more encyclopedic works.[26] In 1381, Raymundus Chamelli de Vivario, also a physician at the Avignon court, wrote a work on the epidemics.[27] Finally, there is an anonymous, undated treatise written by a physician who claimed Johannes de Tornamira as his teacher, which probably reflects late fourteenth-century Montpellier teaching.[28]

While the Montpellier authors at the time of the Black Death implicitly identified the epidemic they witnessed with pestilential fever, albeit recognizing features not described by Avicenna and earlier Montpellier physicians, the late fourteenth-century practitioners were primarily concerned with first constructing a taxonomy of epidemic disease and then placing the current epidemics within their appropriate category. Perhaps because the traditional distinguishing feature of pestilential fever was its etiology, the most important criterion used by the late fourteenth-century practitioners in classifying pestilence was causation: whether the heavenly bodies or putrid fumes arising from sublunary sources were responsible for the epidemics. Also significant in this taxonomic endeavor are the novel features that had been remarked upon by the first generation of Montpellier plague authors, namely, contagion, poison, and apostemes.

Most of the late fourteenth-century authors fit into one of two groups, those who identified the contemporary outbreaks with pestilential fevers, and those who classed them among apostemic diseases. For both groups, the poisonous nature of the epidemics, the contagiousness of the disease, and the appearance of swellings at the emunctories, set the pestilence apart from other similar diseases. The criterion used by both groups to differentiate apostemes accompanied by fever from pestilential fever accompanied by apostemes was whether the celestial cause predominates or the lower cause is involved, respectively.

Jacobi and the anonymous student of Johannes de Tornamira considered the epidemics to be cases of pestilential fever.[29] They differentiated pestilential fever from other fevers in part by emphasizing the poisonous nature of the disease process in the former. Jacobi believed that pestilential fever was a putrid, or humoral fever, but pointed out that it differed from other humoral fevers in that the vapors that attack the heart were more poisonous.[30] The poisonous nature of the vapors provided Jacobi with a mechanism for contagion: the poisonous fumes exuded by the sick individual corrupt the air.[31] Tornamira's pupil agreed that pestilential fever resulted from poisonous vapors, and observed that since the fever was not caused by putrid humors, purgation should be avoided.[32]

As in the pre-plague accounts of pestilential fever and the works composed at the time of the Black Death, discussion of the causes of pestilential fever figured prominently in these two treatises. Instead of adhering to the previous generation's causal chain from the heavens to the admixture of putrid vapors with the ambient air to the corruption and poisoning of the body's moistures and humors, the anonymous student and Johannes Jacobi presented a tripartite taxonomy of the causes of pestilential fever: the heavens alone, the lower causes alone, or both acting in concert. However, with respect to the current epidemic outbreaks, the lower cause was paramount.[33] They agreed with Avicenna and the earlier Montpellier plague authors that the lower cause was noxious sources of fetid vapors, such as marshes and decaying corpses.[34]

Guy de Chauliac and Johannes de Tornamira denied that the plague epidemics were pestilential fevers, and maintained instead that they were some form of apostemic disease.[35] Guy and Johannes de Tornamira agreed with their contemporaries, who did consider the present epidemics to be pestilential fever, that there were three causes of pestilence. While Jacobi and the student of Tornamira believed that the lower cause was the sole, or at least the most important cause of pestilential fever, and accepted the traditional definition of the lower cause as fetid vapors, Guy and Tornamira emphasized the direct action of the heavenly bodies upon the humors and spirits within the body. Guy maintained that the conjunction of the three outer planets in 1345 "impressed such a form in the air and the other elements, that just as the magnet attracts iron, this moved the gross, burning, and poisonous humors and gathered them within where they formed apostemes."[36] These internal apostemes in turn produced fevers, external apostemes at the emunctories, and the spitting of blood. Like the anonymous practitioner, Guy believed that this poisonous disease could be transmitted from person to person by a glance.[37] Johannes de Tornamira presented a similar account of the epidemics, which he called anthracical apostemes (apostemata antrosa). His description of the causes and process of the epidemics parallels Guy's: the heavenly bodies act directly upon the humors, causing them to boil and produce poisonous fumes that make the disease more contagious than pestilential fevers.[38]

One late fourteenth-century author, Raymundus Chamelli de Vivario, classified the epidemics of his day neither as pestilential fever nor as a poisonous aposteme, and instead adumbrated a trend that would reappear in the plague literature of the following century: creating a broader category of pestilential disease and determining what kind of pestilential disease the contemporaneous epidemics were. Raymundus's view of the causes of pestilence most closely resembled that of the late fourteenth-century authors who classified the plague epidemics as apostemes. Like them, he considered the chief cause of pestilence to be the higher cause, and discussed possible mechanisms through which the heavenly bodies acted — through their light, or motion, or influence — concluding that the epidemics were caused by celestial influence.[39] Although Raymundus's etiology was similar to that of the earlier advocates of apostemes, he did not consider pestilence to be an aposteme, but defined a new category — pestilential disease — which encompassed both apostemes. Whereas his contemporary Montpellier authors tended to distinguish pestilence from other diseases because it was poisonous, and to differentiate between pestilential fever and apostemes by their causes, Raymundus used both characteristics to define pestilential disease. In a series of questions, he affirmed that a pestilential disease is one whose efficient cause is the infection and corruption of the air by the heavenly bodies, the lower bodies, or both.[40] Furthermore, he observed that there are three "reasons of pestilential diseases"— malignant air, the invasion of the heart, and the poisoning of the humors — of which the Greek and Arab authors had understood the first two, but only the moderns had recognized the significance of the third.[41] Within his new more general class of pestilential disease, Raymundus placed the two older categories, pestilential fever and poisonous apostemes, naturally characterized by the presence or absence of apostemes.[42]

Although Montpellier appears to have been relatively plague-free for much of the fifteenth century, there were some epidemics at the turn of the century, accompanied by Montpellier additions to the plague literature.[43] The most prominent of the early fifteenth-century Montpellier plague authors is Jacobus Angeli, who served as dean and then chancellor of the faculty, and was the subject of an attack against astrological medicine by Jean Gerson. Angeli discussed *epidimia* in his huge medical encyclopedia, the *Puncta medicine*.[44] Little is known of Michel Boeti beyond what he tells us in his plague treatise: he described himself as a "master of arts and medicine" at Montpellier and Karl Sudhoff identified him with a certain Michel Boel, who wrote his will in 1421.[45] Blasius of Barcelona, who studied at Toulouse and practiced at Montpellier and Sicily, as well as at Toulouse, wrote his treatise in 1406 at the request of King Martin of Aragon.[46] Valescus de Taranta studied at Montpellier while Johannes Jacobi and Johannes de Tornamira were masters, but he may never have taught there.[47] Valescus wrote his plague treatise in 1401 and later incorporated it into his huge compendium of medical practice, the *Philonium*.[48] An anonymous treatise on pestilential fever appears in a codex belonging to the Bibliothèque Nationale.[49] Part of this codex appears to be a collection of medical texts compiled by an unknown student at Montpellier. Immediately preceding the plague treatise, and in the same hand, is a baccalaureate reception speech in which the student thanked his masters Jacobus Angeli and Franciscus Ribauta. This dates the speech between 1412 and 1416, and it is tempting to consider the plague treatise as contemporaneous with the address, and hence representative of ideas about epidemic disease current in Montpellier of that time.[50]

Most of the fifteenth-century Montpellier plague authors continued the taxonomic quest of their late fourteenth-century predecessors. Some classified the epidemic outbreaks as either pestilential fever or as a poisonous aposteme, but most continued in the direction pointed to by Raymundus Chamelli de Vivario: creating a more general category of pestilential disease. However, the fifteenth-century practitioners departed from the late fourteenth-century Montpellier physicians in several respects. First, they considered pestilential disease to be a more general category than the one envisioned by Raymundus, for whom pestilential disease could only be pestilential fever or apostemes. Second, the fundamental distinction between pestilential fever and apostemes for most of the late fourteenth-century authors was etiology; on the other hand, the fifteenth-century Montpellierains emphasized disease process as the criterion for differentiating among pestilential diseases. Finally, the late fifteenth-century authors appear to have been less concerned with distinguishing pestilential disease from other diseases than they were with differentiating between kinds of pestilential disease.

Jacobus Angeli and Valescus de Taranta both wrote about *epidimia*, which they described in a manner reminiscent of accounts of pestilential fever, although they did not use this term. In part, this may be due to the type of works that they wrote. Valescus's *Philonium* is a large compendium of practical medicine, while Jacobus's *Puncta medicine* is an enormous encyclopedia of *questiones* on all areas of medicine. Both works are chiefly compilations and depend heavily upon traditional sources. Valescus, especially, considered many of the questions that we have seen before, such as the different causes of pestilence and how the air is corrupted.[51] Jacobus was more concerned with theoretical and natural philosophical issues than Valescus. He raised questions about which primary element most often generates epidemics, how air putrefies, and the role of the seasons in pestilence.[52]

Blasius of Barcelona followed the fourteenth-century emphasis upon apostemes, asserting that the epidemics of his day are instances of "pestilential glandula."[53] Like the late fourteenth-century physicians who classed the pestilences among apostemic disease, Blasius believed that the heavenly bodies caused the disease directly in some unknown fashion, without the mediation of the elements.[54] However, Blasius also shared characteristics with the fifteenth-century classifiers. He believed that the disease process, namely whether fever preceded the glandula or vice versa, was important in differentiating between diseases. When fever preceded the glandula, the disease was not truly pestilential, but was an ephemeral fever imitating pestilence.[55] Blasius reached this conclusion after his own bout with plague. While he was a medical student at Toulouse, the pestilence returned and he developed an inguinal glandula without fever. His teacher, Philippe Elephantis, cured him by phlebotomy and various medical applications. On the basis of this experience, Blasius concluded that the poisonous matter first was only in the glandula and that the fever was an accident.[56] Furthermore, Blasius briefly mentioned the accidents and symptoms of pestilential glandula, such as quinsy, apoplexy, and paralysis, which would receive greater emphasis in the works by Michel Boeti and the unknown author of the treatise found in the bachelor's manuscript.[57]

Symptoms and accidents of disease were central to the anonymous treatise found in the manuscript of Angeli's and Ribauta's student. This plague treatise is a preventive and curative regimen for pestilential fever. The author did not create a general category of pestilential disease, but devoted nearly one third of the treatise to the treatment of the accidents or symptoms that accompany pestilential fever. The most common of these accidents were malignant apostemes, which he discussed apart from the others.[58] For this anonymous physician, pestilential fever itself had become a broader category, and through a wide range of symptoms encompassed many things.[59]

The notion of a general category of pestilential disease reached its apex in the treatise of Michel Boeti. Like the anonymous fifteenth-century treatise, Michel's was chiefly concerned with the prophylaxis and therapeutics of pestilence, and contained almost no theoretical discussion. The treatise is divided into two parts: prevention of pestilence, and cure of the different *species* of pestilence.[60] Among the different species were pestilential fever and apostemes, glandula or *bocium* in Michel's terminology, but this new category on pestilential disease encompassed more than these traditional forms. Among the other species were deep sleep, delerium, quinsy, pleurisy, variola, and abortion. Michel's species of pestilence overlapped with the accidents mentioned by Blasius and the author of the anonymous text.[61]

For Michel, as for Blasius and the anonymous author, the important characteristic of pestilential disease was the presence of apostemes. Nearly all the species he discussed were a form of aposteme: quinsy was an aposteme of the throat, while glandula appears in the emunctories of the principal members.[62] Nonetheless, some forms of pestilence either did not involve apostemes, or were simply not discussed by Michel. For example, there was no mention of apostemes in the section on pestilential fever.

At the beginning of this essay we remarked upon the important role played by the concepts of balance and imbalance of the humors and primary qualities in medieval explanations of disease. Yet we have seen that in the plague literature emanating from Montpellier, these notions were relatively unimportant. To be sure, fever itself was considered an imbalance in the primary quality of heat, and a superfluity of humors was believed to generate the characteristic apostemes of plague. Furthermore, the traditional notion of maintaining an equilibrium of the humors and qualities was of crucial importance in the prophylactic regimens contained in the plague literature. However, the Montpellier practitioners seemed more concerned with classifying the epidemic outbreaks and attaching specific names to them.

The first generation of Montpellier plague authors, writing at the time of the Black Death, considered the epidemic to be pestilential fever, although they note features hitherto not a part of discussions of pestilential fever: contagion, the poisonous nature of the disease, and the buboes. The second and third groups of Montpellier practitioners writing on pestilence were primarily engaged in constructing a taxonomy of the epidemics. The late fourteenth-century physicians distinguished epidemics from other diseases because pestilence was poisonous and possessed the characteristic aposteme, while they classified epidemics as either pestilential fever or as poisonous apostemes, based primarily on etiology. The fifteenth-century authors were less concerned with differentiating pestilence from other diseases and concentrated upon distinguishing between a large number of pestilential diseases based upon symptoms.

Physicians' experience with the epidemics and their professional role in society are the chief factors in explaining this historical development. Both Guy de Chauliac and Blasius of Barcelona contracted plague themselves and used their personal experience to justify points that they made in their discussions. Furthermore, several of the authors mentioned their professional experiences and observations of the disease, again using these in support of their theories and therapeutic and prophylactic measures.[63] Plague as understood today presents medical practitioners with a complex set of symptoms.[64] There are three types of plague: bubonic, characterized by the appearance of buboes at the lymph nodes; pneumonic, characterized by pulmonary involvement with the spitting of blood as a primary symptom; and septicemic, leading to rapid and certain death. On the basis of the descriptions of symptoms found in medieval plague literature and modern understanding of the natural history of plague, many historians and epidemiologists think that the mid-fourteenth-century outbreaks probably had a high number of pneumonic and septicemic cases, and during the course of the fourteenth century the relative proportion of bubonic cases increased. This may explain the growing importance of apostemes in the discussions of the pestilence and why many of the Montpellier physicians considered the plague outbreaks to be instances of an apostemic disease.

From 1348 on, most of the Montpellier authors emphasized the poisonous nature of the epidemics. The devastating outbreaks were a horrible experience that was difficult to comprehend. The traditional explanatory device of equilibrium appeared inadequate to explain some puzzling phenomena: why the disease affected people other than the most susceptible, why it affected the young and the strong, and the mechanism of contagion.[65] Poison is well-suited to this explanatory role for two reasons. First, poison is an extremely horrible substance that can affect anyone, regardless of a temperate complexion and good regimen.[66] Galen maintained that poisons and venomous creatures were insidious enemies that were difficult to avoid.[67] In a similar fashion, the Montpellier plague authors regarded the epidemics with their poisonous vapors as evil and inimical to humans, even without going as far as Alphonse of Cordova with his malicious poisoners. Second, medieval physicians and natural philosophers believed that poison acts at a distance through occult qualities, which made it a suitable mechanism for contagion.

The intense interest in classification is understandable, if we consider the role of the medical practitioner. Naturally, one of the primary obligations of medieval physicians, was to cure sick individuals. Determining the appropriate therapeutic measures depended upon correct diagnosis of the disease. We have seen that Johannes Jacobi and Raymundus Chamelli de Vivario maintained that without correct understanding of the nature and cause of the epidemics, physicians could not treat their patients properly. In addition, during the fifteenth century (and later) physicians were called upon by local governments to ascertain if sick individuals had plague; if so a quarantine would then be put into effect.[68] Perhaps the Montpellier practitioners were interested in classifying epidemic disease in order to refine diagnosis as the physician's social role during times of pestilence expanded.

The fourteenth- and fifteenth-century practitioners of Montpellier, confronted with a devastating and complex disease, successfully explained this experience within the framework of the medical tradition of their predecessors. In order to explain the new and sometimes anomalous features presented by these epidemics, the Montpellier physicians devised new modes of explanation by bringing disparate facets of medieval medical understanding together to form a coherent whole.

NOTES AND REFERENCES

1. Karl Sudhoff published or described nearly three hundred of these treatises during the first quarter of the twentieth century. An index and statistical summary of this project appears in Karl Sudhoff, "Pestschriften aus dem ersten 150 Jahren nach der Epidemie des 'schwarzen Todes' 1348, XX," Sudhoffs Archiv 17(1925): 264-291.

2. For a discussion of medieval theories of health and disease, based on Italian commentaries on Galen's Tegni, see Per-Gunnar Ottoson, Scholastic Medicine and Philosophy: A Study of Commentaries on Galen's Tegni (ca. 1300-1450) (Uppsala: Institutionen För Ide- och Lardomhistoria, Skrifter Nr. 1, 1982). For a stimulating discussion of these two fundamentally different ways of looking at disease, see Owsei Temkin, "The Scientific Approach to Disease: Specific Entity and Individual Sickness," in A.C. Crombie, ed., Scientific Change (New York: Basic Books, 1963), pp. 629-643.

3. "Volo aliqua de pestilencia scribere que nos frequencius invadit quam fecerit antiquos et ideo pauci antiqui per experienciam ab eis habitam pauca dicere poterunt" (Karl Sudhoff, "Pestschriften, XVIII," Sudhoffs Archiv 17 (1925): 23). Modern scholars disagree over the amount of experience the traditional authorities may have had with epidemic disease. See Manfred Ullman, "Islamic Medicine," Islamic Surveys 12: 86-96 (Edinburgh: Edinburgh University Press, 1978).

4. ". . . factas tempore Ypocras . . . et illam que accidit occupariunt nisi unam partem regionis ista totam mundam" (MS Paris 7133A, fol. 85).

5. Antiqui siquidem in causis talium morborum diminute et superficialiter transiuerunt

quod utique fieri potuerit vel quia rarius et tardius nobis illos viderunt vel quia causas illorum vel modos actionis causarum ignoraverunt quia morbus est incognitus suis non cognitis causis vel modus actionis eorum. Morbus vero incognitus a medico non curatur, ideo antiqui morbos pestiferos non plene curaverunt

(MSS Munich CLM 18444, fols. 200r-v and Vatican City BAV Palat. 1229, fol. 294).

6.   There are two important monographic studies of the plague literature from the time of the Black Death: Anna M. Campbell, *The Black Death and Men of Learning* (New York: Columbia University Press, 1931), and Domenick Palazzotto, "The Black Death and Medicine: A Report and Analysis of the Treatises Written Between 1348 and 1350," (Ph.D. Dissertation, University of Kansas, 1973). Although it deals with a more specialized topic, Darrell Amundsen's, "Medical Deontology and Pestilential Disease in the Late Middle Ages," *Journal of the History of Medicine und Allied Sciences* 32(1977): 403–411 is helpful and treats works written after the Black Death.

7.   On the medical faculty at Montpellier, see Danielle Jacquart, *La milieu médical en France du XIIe au XVe siècle* (Geneva: Librairie Droz, 1981), *passim* and Sonoma Cooper, "The Medical School of Montpellier in the Fourteenth Century," *Annals of Medical History* 2(1930): 164–195.

8.   The frequency of occurrence and severity of plague epidemics seems to have varied tremendously from area to area. See Élisabeth Carpentier, "Autour de la peste noire," *Annales E.S.C.* 17(1962): 1062–1090.

9.   The treatise of the anonymous practitioner is described by Campbell,[6] p. 21 and Palazzotto,[6] pp. 44–45. It was edited by L.A.J. Michon from a corrupt manuscript, (*Documents inédits sur la grande peste de 1348* (Paris: J. B. Ballière et fils, 1860), pp. 71–81); my references will be to the superior recension contained in MS Paris BN 11227, fols. 209–211.

10.   Alphonse of Cordova's treatise is found in Karl Sudhoff, "Epistola et regimen Alphontii Cordubensis de pestilentia," *Sudhoffs Archiv* 3 (1910): 223–226.

11.   Alberti describes the buboes appearing in the emunctories (the "disposal areas" of the principal organs of the body), the classic symptom of bubonic plague, and a symptom not remarked upon in earlier discussions of pestilential fever. See MS Vatican City BAV Palat. 1331, fol. 291v.

12.   The definition contained in Avicenna's *Canon*, the fundamental teaching text of the Middle Ages was the starting point for medieval discussions of fever: "Fever is extraneous heat arising in the heart and proceeding from it, to the entire body by means of the spirits and blood" (Avicenna, *Liber Canonis* (Venice, 1574; reprint Brussels, 1971), Liber IV, Fen I, Tractatus i, Capitulum i). Two important Montpellier masters writing just before the Black Death wrote on fevers, and their discussions closely parallel Avicenna's: Bernard of Gordon in his *Lilium medicinae* (MS Paris BN 11227, fols. 1–203) and Gerard de Solo in his *De febribus* (MS Seville BC 5-1-45, fols. 40–41). On Bernard, see Luke Demaitre, *Doctor Bernard de Gordon: Professor and Practitioner* (Toronto: Pontifical Institute of Mediaeval Studies, 1980). On Gerard see Anne-Sylvie Guenon, "Gérard de Solo, Maître de l'Université de médecine de Montpellier et practicien du XIVe siècle," *Positions des thèses* (École Nationale des Paris, Chartes: 1982), pp. 75–82. There are few modern discussions of medieval theories of fever. Perhaps the most useful recent work on fevers, although focusing on a later period, is Iain M. Lonie, "Fever Pathology in the Sixteenth Century: Tradition and Innovation," W. F. Bynum and V. Nutton, eds., *Theories of Fever From Antiquity to the Enlightenment, Medical History*, Suppl. no. 1 (London: Wellcome Institute for the History of Medicine, 1981), pp. 19–44.

13.   In general discussions of fever, for example in Avicenna's *Canon* and Bernard of Gordon's *Lilium*, the threefold division is discussed, and chapters on the three types of fever follow; a chapter on pestilential fevers follows those chapters, without any explanation of how this fits into the classificatory scheme.

14. Et quando faciunt necessario virtutes agentes celestes, et virtutes patientes terrestres humectationem vehementem aeri, expelluntur vapores, et fumi ad ipsum, et sparguntur in ipso, et putrefaciunt eum cum caliditate debili, et quando fit aer secundum hunc modum, venit ad cor, quare corrumpit complexionem, que est in ipso, et putrefit quod circumdat ipsum de humiditate, et accidit egressa a natura, et spargitur in corporum. (Avicenna, *Liber canonis*, L. IV, F. I, Tr. iv, Cap. 1). For a cogent discussion of *complexio* and its role in late medieval Italian medical discussions, see Per-Gunnar Ottoson.[2]

15. ". . . anno domini mcccxlv facta fuit coniuncti Saturni et Martis et Iovis . . . et quia Mars fuit aliquando retrogradus ideo plus abstraxit de vaporibus a terra et aqua et aerem corrupentibus." (MS Paris BN 11227, fol. 209). This account is not original; it closely follows the Paris faculty's consultation (see Campbell,[6] pp. 39–42).

16. ". . . ascendit inde per venam organiam . . . ad pulmonem et . . . pulmonis motus cessat et non potest ventilare supra cor ad infrigidandum eum et tunc cor calefit et fit febris pestilentialis" (MS Paris BN 11227, fol. 209v).

17. For example, see the discussion of the anonymous practitioner's theory of contagion below.

18. Avicenna did not consider pestilential fever contagious, perhaps because belief in contagion is contrary to Islamic teaching. For the Islamic attitude towards infection, see Manfred Ullman,[3] pp. 86–96. For medieval western attitudes, amassing evidence that certain diseases were considered contagious, see Owsei Temkin, "An Historical Analysis of the Concept of Infection," *Studies in Intellectual History* (Baltimore: The Johns Hopkins University Press, 1953), pp. 123–147; Charles and Dorothea Singer, "The Scientific Position of Girolamo Fracastoro 1478?–1553 with Especial Reference to the Sources, Character and Influence of His Theory of Infection," *Annals of Medical History* 1 (1917): 1–34; Mirko Grmek, "Le concept d'infection dans l'antiquité et au Moyen Age, les anciennes mesures sociales contre les maladies contagieuses et la fondation de la première quarantaine à Dubrovnik (1377)," *Rad. Jugoslavenska Akademija Znanosti i Umjetrosti* 384 (1980): 9–54; and Vivian Nutton, "The Seeds of Disease: An Explanation of Contagion and Infection From the Greeks to the Renaissance," *Medical History* 27(1983): 1–34.

19. Avicenna, *Liber Canonis*, L. IV, F. I, Tr. iv, Cap. 13, contains Avicenna's discussion of fevers and apostemes.

20. There are other forms of plague besides the bubonic. See below.

21. For example, Bernard Alberti writes:
    . . . si sint humores superflui in epate qui mittantur ad inguina vel in cerebro mittantur retro aures et si in corde ad acellas vel ad alia membra . . . ex quibus humoribus sit dispositus causabuntur bubones, herisipili, carbunculi, et alia diversa genera apostematum.
(MS Vatican City BAV Palat. 1331, fol. 138v).

22. ". . . et aliquando cerebrum expellit hanc ventosa et venenosam materiam per nervos opticos concavos ad oculos . . . et quem spiritum visibilem si quis sanus aspexerit suscipit impressionem morbi pestilenciales." (Paris, BN MS 11227, fols. 209v–210).

23. Exemplum de basilisco qui respexerit fortiter aliquem sanum ipsum respicientem statim spiritus visibilis et aereus et venenosus egrediens ab oculi basilisci transiens in obiecto, scilicet in oculis respicientis basiliscum statim et subito intoxicat predictum hominem . . . legitur in libro Aristotelis ad Alexandrum [that is, the *Secreta secretorum*] de quadam domicella nutrita ex veneno per quamdam reginam quam illa regina misit Alexandro ut occidit Alexandrum solo viso et concubito suo quam domicellam videns Aristotelem novit per oculos suos ipsam esse venenosam
(MS Paris BN 11227, fol. 210).

24. "Experientia docuit quod ista pestilentia non vadit ex constellatione aliqua et per conse-

quens nullam naturalem infectionem elementorum, sed vadit ex profundo malitiae per artificium." (Sudhoff,[10] p. 224).

25. For biographical information on Guy, see Ernest Wickersheimer, *Dictionnaire biographique des médecins en France au Moyen Âge*, 2 vols., (Geneva: Librairie Droz, 1936; reprint Geneva, 1979), pp. 214–215. Guy's account of the pestilence has been translated into English by Michael McVaugh, in Edward Grant, ed. *A Sourcebook in Medieval Science* (Cambridge, MA: Harvard University Press, 1974), pp. 773–774.

26. For biographical details, see Wickersheimer,[25] pp. 422–424 (for Jacobi) and p. 494 (for Tornamira). The dispute over the chancellorship is recorded in the cartulary of the university: *Cartulaire de l'Université de Montpellier*, 2 vols., (Montpellier: Ricaud, 1890), I: 460–461.

27. For biographical details, see Wickersheimer,[25] p. 674. Raymundus's connection to Montpellier is not absolutely certain. Some support is found in the manuscript tradition; the end of one manuscript version of his treatises says "Compositus per me Raymundum Chamelli magistrum in artibus et in medicine preclari Montispessulani" (MS Munich CLM 18444, fol. 266).

28. The treatise by Tornamira's pupil is problematic. It is found in a unique manuscript (MS Oxford Bodl. Canon. misc. 524) written in a late fifteenth-century hand. Many references to individuals and institutions contained in the text are confused, but these may be due to scribal errors. One of the difficult passages is the one in which the author claims to be a student of Tornamira's: "et dixit dominus meus magister Johannes de Tornamira, compilator Clarificatori et cancellari predicti studii parisiensis" (fol. 105v). Tornamira did write the *Clarificatorium*, a commentary on the ninth book of Rhazes's *Almansor*, but was the chancellor of the Montpellier, not the Parisian, medical faculty. Despite the problems, I am inclined to accept the Montpellier provenance of the work.

29. Johannes, indeed, criticizes those physicians who do not accept that epidemics caused by the lower cause alone are pestilential fevers: "inde potest provenire febris pestilenciales multi medici sunt decepti non credentes talem febrem esse pestilencialem et ideo male sciunt curare" (Sudhoff,[3] p. 24).

30. Febris autem pestilencialis semper est continua et humoralis putrida et bene potest exacerbare. Differencia autem inter ipsam et alias febres putridas es quia pestilencia fumi putridi venientes ad cor de minera sunt magis venenosi

(Sudhoff,[3] pp. 24–25). Jacobi's account in his plague treatise agrees with his discussion of pestilential fever in his *Secreta practice*, a compendium of practice written for Charles VI of France: "sed in febre pestilenciali tales fumi sunt putredinales et veneosi per admixtionem aeris corrupti" (MS Seville BC 5-1-45, fol. 108).

31. "Ad secundum questionem: si tales morbi sint contagiosi, dico quod sit quia a talibus corporinus effumant fumi venenosi corrumpenter aerem et humiditates" (Sudhoff,[3] p. 25).

32. "Nota quod in hac febre non confert purgacionibus in sistere quia humor putredus non est causa sed vapor aeris venenosus et putredus cor inficius et cuius humiditatem" (MS Oxford Bodl. Canon. misc. 524, fol. 103v).

33. Much of the prophylactic advice proffered involves avoiding the sources of noxious fumes and rectifying the air.

34. Tornamira's student: ". . . a radice inferiori, scilicet a paludibus vel cadaveribus seu fetoribus et huiusmodi" (MS Oxford Bodl. Canon. misc. 524, fol. 103). Jacobi:
A radice inferiori venit solum ut nos videmus, ex latrine propre cameram vel ex alia re particulari . . . Quandoque autem venit pestilencia a re inferiori multum magna et tunc potest eciam dici pestilencia universalis, ut accidit ex cadaveribus mortuorum ut post bellum

(Sudhoff,[3] pp. 23–24).

35. For example, Johannes writes: "Cum morbi pestilenciales qui hactenus fuerunt, sunt et erunt . . . sint communiter apostema antrosa . . . ad quae sequitur febris ipsa concomitans, non quod ipsa febris sit pestilencialis" (Karl Sudhoff, "Pestschriften III," *Sudhoffs Archiv* 5 (1912): 48).

36. "Impressit enim talem formam in aere et in aliis elementis quod sicud adamas attrahit ferrum ita ipsa movebat humores grossos, adhustos et venenosos et congrebat eos ad interiorem qua et faciebant apostema" (MS Paris BN 7133A, fol. 85v). This fifteenth-century copy of Guy's *Magna chiurgia* belonged to Jacques Angeli, who will be discussed below.

37. ". . . et fit tante contagiositatis specialiter que fit cum sputo sanguinis quare non solum morando sed etiam inspiciendo unus respiciebat de alio in tantum" (MS Paris BN 7133A, fol. 84v).

38. ". . . nam talium fiunt potissime ex radice superiori, quia ex aspectu orbis facientis ebullicionem, combustionem et putrefactionem . . . nam ab istis fumus horribilis venenosus et pestiferus elevatur ad cor per arterias et venas et alios meatus, propter quam effumationem talia apostemata sunt magis contagiosa quam febres pestilenciales" (Sudhoff[35]).

39. Ymmo ipso aere sereno existente et apparente tranquillo puro frigido et monstruoso fortior est ibi ipidimia et sic maior ut patet hodie et in magnis humidis corporibus et fortibus ut in pueris et iuvenibus et hoc est signum forte quid est ab influencia celi que fit insensibiliter

(MS Munich CLM 18444, fol. 225).

40. ". . . omnis morbus dicitur pestilencialibus cuius causa efficiens est aer alteratus, infectus et corruptus a causa superiora, inferiori vel utrumque." (MS Munich CLM 18444, fol. 245v).

41. De ratione morbis pestilencialis sunt tria: primum aeris malignatio, secundum cordis prima invasio, tertium humorum invenenatio. Primas duas condiciones posuerunt antiqui doctores scientie fundatores ut Ypocras, Galienus, Rasis et Avicenna in locis sepe allegatis. Tertium posuerunt moderni doctores ut magnus magister Gentilis de Furlino et magister Jacobus Rotunda Romanus

(MS Munich CLM 18444, fol. 245v).

42. This division is clearly seen in the structure of the therapeutic portion of his treatise: Tertia pars principalis huius tractatus est de cura morborum pestilentialium magis particulosorum et habet duo capitula: primum est de apostematibus venenosis simul cum febres pestiferis existentibus, secundum est de febribus pestilentialibus sine apostematibus

(MS Munich CLM 18444, fol. 245v).

43. Louis Dulieu, *La médecine à Montpellier*, t. 1 (Avignon: Les Presses Universelles, 1975), p. 181, mentions an epidemic in 1407 and then none until 1481. The town chronicle, *Le Petit Thalamus* (Montpellier: Martel, 1836) however, mentions pestilences or mortalities occurring in 1407, 1408, 1409, and 1411; of course all of these outbreaks need not be plague.

44. For biographical details, see Bruno Delmas, "Le chancelier Jacques Angeli et la médecine à Montpellier au milieu du XVe siècle," *Positions des Thèses* (Paris: École Nationale des Chartes, 1966), pp. 23–28. As far as I know the only reference to Angeli's deanship is found in an anonymous baccalaureate address (see note 50 below). The *Puncta medicine* exists in three codices of Seville: MSS Seville BC 5-7-16, 5-7-17, and 5-7-18.

45. Sudhoff,[3] p. 46.

46. Sudhoff,[3] pp. 105, 113. and MS London Sloane 428, fols. 145v, 151v.

47. For biographical details, see Wickersheimer,[25] p. 772.

48. In addition to the copies found in the *Philonium*, there are separate manuscript versions: MSS New York NYAM 4, fols. 294–301v; Paris BN fr. 630, fols. 43v–49; Seville BC 5-1-45, fols. 226–226, 259v–262. The Seville version was actually copied from a copy of the *Philonium*.

49. MS Paris BN nal 3035, fols. 69v–80v.

50. ". . . magistro meo Jacobo Angeli in artibus et medicine digno professori atque decano huius

facultatis medicine . . et medicine professori" (MS Paris BN nal 3035, fol. 69v). We know that Angeli left Paris in 1412 and then presumably studied medicine; the next date we have for him is receiving his licentiate in 1417 at Montpellier (Delmas,[44] p. 24). Likewise, we know that Ribauta left Montpellier in 1416 (Wickersheimer,[25] p. 157).

51. Valescus de Taranta, *Practica que alia philonium dicatur* (Venice, 1502), fols. 189–191.

52. MS Seville BC 5-7-16, fols. 152v–155v.

53. ". . . loqui de glandulis pestilentialibus et ipsa epydimia nam talis pestilentia hoc nostro currit tempore" (MS London BL Sloane 428, fol. 145v and Sudhoff,[3] p. 103).

54. ". . . scimus itaque pestilentiam glandulam ab asteris causata, sed a quibus et quomodo ignoramus, illud autem scimus, quia non fertur vento nec surgit a terre visceribus, ergo a superioribus, sed non ab aere" (MS London BL Sloane 428, fol. 146v and Sudhoff,[3] pp. 106–107).

55. "Dico quod cum febris precedit glandulam, illa febris non est epidimialis sed effimera que superveniente glandula imitatur in pestilenciam" (MS London BL Sloane 428, fol. 150; missing in Sudhoff's edition).

56. Since this portion of the treatise is garbled in Sudhoff's edition, it is worthwhile quoting *in extenso.*

> Ego vero studebam Tolose redit pestilencia que iam cessaverat duobus annis et peremit ultra cc studentes et paucos cives. Et tunc super hoc venit michi glandula in inguine dextro sine febre, quare stati aperui sophenam dextri pedis. Et nocte sequenti supervenit alia glandula in inguine et prima cum dolore forti crevit et cum hoc me febre fortiter invasit quare timens et surgens mane, feci venire meum Phylippum et lephantis, magnum et universalem phylosophum et valentem medicum ut interesset maioris glandule apertioni in principio reunebat. Sed meis rationibus quia materia erat furiosa et morbus acutissimus oportebat eam evacuari presto per locum propinquissium aquievit, quare imposito flebothomio per digit spacium in profundum et tantumdem in latum exivit parum sanguinis, et tantum de aqua crocea quam tamen capi poterit in testa avellane, qua emissa statim febre evanuit supra cuius apertionem. Magister meus apposuit medicinam glutinarem nec ob hoc destiti a ergerem febrem tunc vero nocte sequenti iuvenili farmaco evacuari quem quies et surgens mane ammovi emplastrum et reperi vulnus consolidatum taliter quod vix apparebat cicatrix. Et sic fui curatus mirante magitsro meo curam tam subito terminari. Ex quo experimento tria precepi: primum quod materia venenosa in principio invasionis non erat nisi in glandula, secunda febris fuit accidens eius, tertia quod tardato in evacuando est facere infirmum periclitarem

(MS London BL Sloane 428, fols. 151v–152).

57. "quedam pestilentia sit cum squinancia alia, cum apoplexia alia, et alia cum paralisi, alia cum faringis, alia cum emigrana, alia cum pleurisi, alia cum sputo sanguinis, alia cum causone, alia cum disinteria" (MS London BL Sloane 428, fol. 148v and Sudhoff,[3] pp. 108–110).

58. "Et primum quidem accidens et magis generale seu accidentia magis generalia sunt struma seu glandula vel bubo et flegmones antraces et similia apostemate maligna" (MS Paris BN nal 3035, fol. 74v).

59. There are twenty-seven accidents or symptoms: dolor capitis, frenesis, subecli vel litargia, vigilie, fluxus sanguinis narium, sitis, nigredo lingue, excoriatio oris et trachee artherie, squinancia, tussis, pleuresis, peripulmonia, sputus sanguinis, sincopis, vomitus, dolor stomachi, vermes, dolor et calor epatis, dolor splenis, dolor ventris, yliaca passio, fluxus ventris, restrictio ventris, retencio urine, variole et morbilli, and sudor dyaforeticus (MS Paris BN nal 3035, fol. 75v).

60. "Regimen pestilencie dividitur in duo, scilicet preseruativum et curativum. Curativum in fine ponentur, quando particulariter tractabitur de qualibet specie pestilencie" (Sudhoff,[3] pp. 46–47).

61.   The different chapters are: somnus profundus, frenesis, glandula, squinancia, antrace, bocio et glandula, febris pestilencialis, pleuesis, fluxu ventris, heresipela, pustula ignea, variole et morbillo, and aborsu pestilenciali (Sudhoff,[3] pp. 48-49, and MS Leipzig KMU 1162, fols. 356-359v). Chapters 3 through 9 do not appear in the MS; either they are missing or the chapters are incorrectly numbered.

62.   "Alia est species pestis squinancia: est apostema gutturis" (MS Leipzig KMU 1162, fol. 356v. This is missing in Sudhoff's extracts). "Alia species pestis est que vocatur glandula seu bocium que fit in emunctorio membri principalis" (fol. 356v).

63.   For example, Johannes de Tornamira described a successful phlebotomy to support his opinions on the matter (Sudhoff,[35] p. 53).

64.   The modern literature on plague is vast. The following are recent accounts, and also contain pointers to the literature: Jean-Noël Biraben, *Les hommes et la peste en France et dans les pays européens et méditerranéens*, 2 vols., Civilisations et Societies, ser. 35-36 (Paris and The Hague: Mouton, 1975); *The Plague Reconsidered: A New Look at Its Origins and Effects in 16th and 17th Century England* (Stafford: Local Population Studies, 1977); Robert S. Gottfried, *The Black Death* (New York: Basic Books, 1983).

65.   Several of the authors discussed above demonstrate how poison allows a satisfactory explanation of contagion. In note 39, Raymundus Chamelli de Vivario's claim that plague attacked the young and the strong is given. Valescus de Taranta maintained that the disease first attacks the "prepared," namely the most susceptible, but then attacks others because it is contagious ("quia contagiosus morbus est alios cuiuscumque complexionis fuerint potest inficere cum ab eis fumi mali et corrupti et venenosi," *Philonium*, fol. 189).

66.   "Venenum est materia horribilissime qualitatis" (Johannes de Tornamira, *De febribus*, MS Bern 570, fol. 100v(59v)).

67.   Cited in Gilbert Watson, *Theriac and Mithridatium: A Study in Therapeutics* (London: The Wellcome Historical Medical Library, 1966), p. 1.

68.   On Montpellier plague physicians and surgeons of the late fifteenth century, see Dulieu,[43] pp. 178-184. For an early documented example of quarantine laws, see Grmek,[18] pp. 35-54.

Science for Undergraduates in Medieval Universities

EDITH DUDLEY SYLLA

History Department
North Carolina State University
Raleigh, North Carolina 27695-8108

A S PART OF THE larger problem of trying to understand how medieval science was shaped by its social, institutional, and intellectual context, I want in this paper to direct attention to those scientific works written in connection with arts education in medieval universities. Of course, not all medieval science was done in universities. Some medieval science was done for kings, emperors, or other secular patrons. For instance, the work *On the Art of Hunting with Birds* was written by the Emperor Frederick II.[1]

Before the rise of universities, some science was done in connection with cathedral schools. This would include, for instance, the work of Bernardus Silvestris.[2] Unfortunately, we know very little about the life of Bernardus Silvestris, but we know that his *Cosmographia* was dedicated to Thierry of Chartres.[3] We know that the rhetorician Matthew of Vendome was Bernard's student, probably at Tours.[4] The *Cosmographia* was recited to Pope Eugene II in 1147.[5] The *Cosmographia* itself reveals familiarity with recent work in science, including translations from the Arabic, but it is written as a sort of epic poem. Thus we should probably consider Bernard as a poet and rhetorician, but he belongs to a period before scholasticism had set in very strongly in the schools. In this earlier period there was, in Winthrop Wetherbee's words, a "Platonist synthesis of learning and expression as a means of philosophical and religious understanding."[6] Shortly after Bernard there was an increasing differentiation between literature, on the one hand, and the sciences of the quadrivium and Aristotelian philosophy on the other. A work like the *Cosmographia* would not likely have been written in the later university context.

I would like to understand better than I do who wrote on alchemy in the Middle Ages, what their institutional context was, and what their intended audience. Although works on alchemy are ascribed to university masters such as Albertus Magnus and Thomas Aquinas, I do not believe that universities were the context in which alchemy was done—at the very least I do not believe it was ever an official part of the curriculum.

I would also like to understand better the institutional context of astrology — this I think sometimes *was* done by university masters even if it was not part of the official curriculum (that is if judicial astrology is distinguished from what we would call astronomy), but there is the complication that the word *astrologia* was often used in medieval works for both astronomy and astrology. Another complication arises from the use of astrology of a sort, that is consideration of the positions of the planets, in medicine. Hence at Padua astrology apparently *was* part of the curriculum and was taught by physicians.[7]

So my topic in this paper is meant to be limited to that sort of science done in the intellectual context of universities. In my title I have limited my topic to science for undergraduates. This was meant to set aside, for the moment, the teaching of science in graduate medical faculties. It also sets aside the use of science in theology, as this would be connected with the work of graduate faculties. I do not, however, mean to set aside the work that fourteenth-century students did between "determining" (about the time they became bachelors of arts) and "incepting" (about the time they became masters of arts). In other words, I will take as my subject all the science that arts students did before becoming masters of arts. (This grouping is natural especially because, as was pointed out in the discussion at the Barnard conference, medieval universities often did not make a sharp distinction between pre- and post-baccalaureate study in arts.) I have put the limitation "fourteenth century" into my title just because that is the period I know best, but because of the nature of the data I have used, what I say probably applies just as well to the late thirteenth and early fifteenth centuries at least.

What then was the science for arts students in fourteenth-century universities? Oxford is the case with which I am most familiar. Among fourteenth-century universities in the north (excluding, that is, Italian universities where there was often a joint faculty of arts and medicine), Oxford has the reputation of being the strongest in the sciences, but I think its science teaching likely differed at most only in degree and not in kind from science teaching elsewhere. The two main medieval Oxford documents concerning lecture requirements before determination come from 1268 and 1409. In 1268 before determining the student was to hear three books of natural philosophy, namely Aristotle's *Physics*, *De Anima*, and *De generatione et corruptione*.[8] In 1409 before determining, the student was to hear the *Algorismus integrorum*, the *Compotus ecclesiasticum*, and the *Tractatus de Sphaera*.[9]

This chronological change goes somewhat counter to what one might expect from the statutes of the University of Paris, where the subjects of the quadrivium were mentioned as early as 1215, while the *De Anima* first appeared in 1252, and other Aristotelian books first appeared as subjects of

masters' lectures in 1255.[10] Should we conclude that according to the statutes
of 1268 Oxford students did not study the subjects of the quadrivium before
determining? Probably we should not conclude this, but only that the books
of the quadrivium were not thought particularly relevant to the student's read-
iness to determine, in which activity he would dispute mainly concerning logical issues or more rarely concerning grammar or other subjects. But more
on this later.

From a statute from before 1350 at Oxford, we learn that before incepting
as a master of arts the student needed to have heard, in addition to the books
required for determination, Aristotle's *Politics* or his books *De progressu et
de motu animalium*, the other books on natural philosophy, and the *Ethics*.
Furthermore, whether the student had determined or not, he should have heard
six books of Euclid, Boethius's *Arithmetic,* and the *Computus, Algorismus,*
and *De sphaera* mentioned in the 1409 statutes.[11] In 1431 before incepting
as master the student should have heard the complete course of study in the
seven liberal arts and in the three philosophies, that is natural philosophy,
ethics, and metaphysics. Specifically in the sciences, he should have heard
Boethius's *Arithmetic,* Boethius on music, and in geometry Euclid's geometry
or Alhazen or Witelo on perspective, the *Theorica Planetarum* or Ptolemy's
Almagest. In natural philosophy he should have heard the *Physics, On the
Heavens and Earth*, on the properties of elements, or on meteorology or on
vegetables and plants, or *De anima* or *On animals* or some of the *Parva
naturalia*.[12]

It is clear, then, that throughout the later medieval period the student was
expected to study the quadrivium and natural philosophy before becoming
a master of arts. If there is any difference between the earlier and later statutes, it may concern the order in which these subjects were studied or whether
given subjects were to be studied before or after determining.

At Paris a similar situation seems to have prevailed. According to the Paris
reforms of 1366, the student should have studied before licensing Aristotle's
De anima, Physics, On generation and corruption, On the heavens and world,
and the *Parva naturalia*. They should also have studied certain mathematical
books (*aliquos libros mathematicos*) not further specified. Before incepting,
the student should have heard further the *Meteorology*.[13]

From these statutes we can conclude, then, that at least at certain periods
during the fourteenth century, arts students at Oxford and Paris — and something similar could be shown for other universities — were required to attend
lectures on the quadrivium and on many of Aristotle's works of natural philosophy. We know, moreover, that fourteenth-century arts students did not
hear only the lectures required *pro forma*. Lectures were given and heard on

a number of other scientific works. Not rarely, students could obtain dispensations to count attendance at such a nonprescribed series of lectures in place of lectures on one of the books prescribed *pro forma*. On other occasions students seem to have heard such nonprescribed lectures simply for their intrinsic interest—as "electives" in a system where there was no such thing as credit for elective courses. There are also regulations which state that on holidays—or particularly on the afternoon of days preceding holidays—extraordinary lectures on certain of the scientific books could be continued. Whether this means that such lectures were particularly in demand or that, being less important, they could go on when full attendance or attention could not be expected is not entirely clear.[14]

If these statutes represent the normal context of science for arts students in medieval universities, is it fair to assume that most or all fourteenth-century editions of works on the quadrivium and most or all medieval commentaries on Aristotle's physical works were produced in connection with this university arts teaching? Surely not *all* such works can be located in a university context. When Nicole Oresme wrote his commentary on Aristotle's *On the Heavens* in French, obviously he meant it for King Charles V and the King's circle of advisors and not for university students.[15] Oresme ended his book, however, with a paragraph saying:

> Thus, with God's help, I have finished the book on *The Heavens and the World* at the command of the very excellent Prince Charles, the fifth of this name, by the grace of God, King of France, who, while I was doing this, has made me Bishop of Lisieux. For the purpose of animating, exciting, and moving the hearts of those young men who have subtle and noble talents and the desire for knowledge to prepare themselves to argue against and to correct me because of their love and affection for the truth, I venture to say, and feel quite certain that no mortal man has ever seen a finer book of natural philosophy in Hebrew, in Greek, or Arabic, in Latin or French than this one.[16]

This sounds very much as if he is thinking of the young men as students.

Similarly, when Walter Burley dedicated some of his Aristotelian commentaries to Richard de Bury, Bishop of Durham, and later to Pope Clement VI, it is not clear how thoroughly he revised works that had had university origins.[17] In the All Souls College, Oxford, manuscript of Burley's *Physics* there are two dedicatory letters, one prefacing the whole work and the other prefacing Books 7 and 8.[18] The opening sentence of each letter is a formal dedication to the masters and scholars of Paris, but each goes on to eulogize Richard de Bury as Burley's patron. Burley states that a long time ago while at Paris, he had commented on the *Physics*, Books 1–6, and that Bury had asked him to complete his study by writing on Books 7–8. He says he plans to use the same mode and order in commenting on the later books. Bury, he says, directs "his clerks to concentrate their efforts on the fields in which they have special competence and to put in writing the more important and useful results of their researches."[19]

A closer examination and comparison needs to be made, then, between the various versions of Burley's commentaries to understand whether works written in the university context would differ from works written for a bishop patron. We know that Burley, along with other scholars, was engaged as one of a circle of intellectuals assembled by Bury.[20] How did this group function? For those scholars who had a university teaching career in the arts and later went on to other church careers, it is often assumed in default of any concrete evidence that their scientific works were written as part of their university work in arts, but, as we have seen, this is not always a safe assumption.

A third case in point on this subject may be drawn from the life and works of Richard of Wallingford. Wallingford spent two periods in study at Oxford, the first from *ca.* 1308 to *ca.* 1314, when he determined in arts, and the second from *ca.* 1317 to 1327, during which time he presumably incepted in arts and progressed to the status of a bachelor of theology.[21] He spent the rest of his life, from 1327 to 1336, as Abbot of Saint Albans. In his recent edition of the works of Wallingford, John North dates most of Wallingford's astronomical and astrological works, namely the *Canones, Tractatus, Exafrenon, Albion,* and *Rectangulus,* to his second Oxford period, when Wallingford was a bachelor or higher of arts and a student of theology. He dates his treatises on his astronomical clock, his *De sectore,* and possibly his *Kalendarium* to the years when he was Abbot of Saint Albans, during which years he promoted the construction of an astronomical clock for the abbey.[22]

Now Wallingford was unusual and probably the best astronomer in England at the time. Moreover, he was regarded as having perhaps spent too much time on scientific studies. In his history of the Abbots of Saint Albans, Thomas Walsingham wrote:

> We have often heard him complain, not without a sigh, on two scores. First, he left the cloister for study far too quickly and at too early an age [by North's calculation he was about 16 years old] and even before his training was complete. Second, that leaving out other philosophical studies, he gave great attention (in fact more than was fitting) to mathematical pursuits, in which he was particularly learned, namely to speculation on the conclusions of arithmetic and geometry, astronomy and music. His writings and the instruments he made (*opera instrumentorum*) testify to this. His attention being thus distracted, he

studied theology and the rest of philosophy all the less—a fact which caused him grief. He was, however, in the judgment of the Masters, not so incapable of opposition and responsion in the Schools as to be unworthy of the status ' of a Bachelor and of lecturing on the *Sentences*.[23]

About Richard's clock, the *Gesta abbatum* reports:

> [Richard] constructed a horologe in the church, and that he did nobly, and with great expenditure of money and industry. He did not abandon the work as the result of its disparagement by the brethren, although they, wise in their own eyes, regarded it as the height of folly. . . . Indeed, when on some occasion the illustrious King Edward, the Third from the Conquest, had come to the monastery to pray, and saw that so sumptuous a work had been put in hand, while the church was still not rebuilt after the ruin which had come about in Abbot Hugh's time, he modestly upbraided Abbot Richard, on the score that he had neglected the fabric of the church, and spent so much on a less important work, namely the aforementioned horologe. . . . [24]

As North remarks, the construction of Wallingford's astronomical clock required "mathematical and astronomical training going well beyond the limits set by the ordinary curriculum for the *quadrivium*."[25] About the mathematical treatises, however, North says, "In mathematics his merit was to assemble existing knowledge, to organize it in a way which brought out the best of the scholastic method, and to make it accessible to the universities in the form of convenient treatises. After perhaps a century of use by a few Oxford mathematicians the works were generally lost to view. . . . "[26]

Wallingford seems, then, to be an exception proving the rule that most fourteenth century quadrivial writing was done in connection with university teaching. Although Wallingford's astronomical *activity* clearly took place at Saint Albans as well as at Oxford, North places the audience for Wallingford's astronomical *writings* at Oxford, with perhaps some indication of a different audience for some of Wallingford's astrological work.

Wallingford's mathematical and astronomical knowledge may have been at the upper limits for his time and place, and one might justifiably wonder whether his intent was that his works should be used in *teaching* or whether they were not written rather for scholar-astronomers in the Oxford community, for instance for John Maudith, who later belonged to Bury's circle, or for other astronomers at Merton College, to be used in some sort of astronomical activity going beyond teaching as such. On the model of modern science where there is a great mass of specialized scientific writing intended not for "students" as such, but for experts or researchers in the field, we might assume without thinking that the intended audience for medieval scientific works was likewise an audience of other experts. My sense regarding Wallingford is that he may indeed have written at a level beyond the comprehension of most students. On the other hand, his reported discomfort at having devoted such a great proportion of his university time to mathematics and the reported skepticism of the monks of Saint Albans and of the King about the wisdom of building so expensive a clock at Saint Albans, both testify to Wallingford's status as an outlier. He did not quite fit the rule of science gaining its usefulness from its connection with university teaching, but this put him in a somewhat uncomfortable position.

I have two other bits of evidence relevant to the question whether most later medieval work on the quadrivium belonged to the context of university arts teaching. First, we might consider the nature of medieval Latin translations, adaptations, or commentaries on Euclid's *Elements*. Some of the more important of these translations and adaptations have been characterized by the fact that they had frequent additions of a didactic nature – comments made to help the student understand the basis of Euclid's proofs or the logical structure of the work and so forth.[27] The situation for Euclid's *Elements* may be representative of that of other quadrivial works as well. Second, there is the question of so-called practical arithmetics or practical geometries. In his recent edition of one such work, Stephen Victor has concluded that practical geometries were not so much used to teach students how to measure the height of a tower at a distance, as they were simply books to teach mathematics, with the practical applications to be used simply as student exercises, like "story problems" in modern textbooks of elementary mathematics.[28] After extensive study of the *Quadrans vetus*, on the other hand – a work describing how to make and use the instrument of that name – Nan Hahn concluded that we have little basis for concluding whether the book had its greatest use simply as a textbook exercizing mathematical and astronomical skills or whether its major use was as a manual teaching how to construct a quadrans vetus for actual practical use.[29] Few physical exemplars of the instrument itself are extant, but this could equally well be the result of the once-existing instruments being worn out with use as an indication that few instruments ever existed. If instruments *were* constructed, they might have been constructed for teaching purposes as well as for use in extra-university practice.

In my mind, then, the question whether a university teaching purpose can be assumed for most later medieval works on the quadrivium and natural philosophy does not have a simple yes or no answer. Perhaps, indeed, it might be thought unrealistic to expect that such a question will have a clear answer. In the fourteenth century as in all other times including the present, it might be argued, people are often unclear, if not downright confused, about their own motives and intentions. Medieval authors, themselves, might not have made a distinction between writing a book for teaching purposes and writing a book for use by other masters or intellectuals in general for nonteaching purposes.

Why, indeed, did the medieval university curriculum include the quadrivium and natural philosophy? Nowadays one often finds a proliferation of elementary science and mathematics courses in universities, with separate elementary physics courses, for instance, for engineering students, for physics and other science students, and for "poets," *i.e.* for nonscience majors. Was the medieval undergraduate's need for or attitude toward his mathematics and science courses most like that of the modern engineering student, the physics student, or the nonscience major? Presumably, he was like the engineer in that he might expect to apply what he learned in science courses in his future work, albeit the application might be in theology rather than in industry. Presumably too, however, the medieval arts student was like the modern science major in the fact that the best such students could expect to equal or exceed their teachers in knowledge and eventually to become teachers of the subject themselves. The modern nonscience major is required to take science courses to learn a bit about the scientific method, to gain some knowledge about the natural world, and with the hope that he or she will be able to cope with social problems that may arise involving science and technology. In the Middle Ages problems of technology and society were by no means as prominent as they are now, but I would suppose that the medieval undergraduate otherwise had goals and motivations in studying science matching those of the modern nonscience major.

If such broad and, for the moment, unsubstantiated claims about the motivations for the medieval arts student to study mathematics and science are acceptable, then they lead to the following point: the medieval student could have such combined goals because for him the methods of the science or philosophy student, the methods of the post-graduate "scientist" or philosopher, and the methods of the theologian were all the same or similar and not different as they are today. What the medieval student and master, philosopher and theologian all needed to know was how to understand, interpret, compare, and organize textual material, how to expound and explain this material both orally and in writing, and how to come to a determination on disputed points, stating conclusions and supporting them with arguments and evidence and refuting contrary opinions. The student learned these methods by observing the masters using the methods and by using them themselves. (Indeed, one of the points that emerged in the discussion at the conference at Barnard was that in medieval discussions of the theory and practice of a science like medicine, the practice was often taken to be teaching. As in any other medieval guild, what the apprentice in the university may have seen himself as learning was how to function as his master functioned, and, in this case, the master's function may have been seen as that of teaching arts subjects.)

In light of this similarity of method, therefore, one should not be surprised that it is hard to distinguish in medieval natural philosophical works between what was written for students and what was written for other experts. As we use the term "scholar" to refer both to young students and to the most eminent experts, so the term student or *studens* as applied to the medieval context cannot be limited to pupils enrolled in an educational program, but comprehends all those applying themselves to the understanding of various subject matters.

I conclude, then, that most fourteenth-century editions of works on the quadrivium and most medieval commentaries on Aristotle's physical works may well have been produced in connection with university arts teaching, but that the same works probably also were suitable for use by masters or experts. Moreover, at least for natural philosophy if not for the quadrivium, the process of teaching was not sharply distinguishable from the process of doing science. In both cases the scholastic method of raising and replying to questions was the predominant activity.

With this point of view in mind, then, I want to raise questions about some of the main genres of medieval scholastic literature on science. My first question concerns Aristotelian commentaries and related natural philosophical works. In a stimulating article on the longevity of the Aristotelian world view, Edward Grant has argued that the Aristotelian world view lasted as long as it did in part because the commentary form was unsuitable for bringing about revolutionary change.[30] The various changes and improvements that medieval commentators suggested concerning Aristotelian physics lay dormant, in Grant's view, because they remained isolated and uncoordinated as ad hoc solutions to separate problems raised by the Aristotelian text and because they were never brought together to form a new alternative paradigm. A similar point was earlier made by Ernest Moody in arguing that the fourteenth-century physics that Pierre Duhem saw as a precursor of the physics of Galileo never existed as a coherent entity.[31]

Now, of course, it is right that when scientific work appeared in the form of commentaries on Aristotle one got many separate treatments of separate subjects with the order often determined by the chance order of Aristotle's casual remarks rather than by any intrinsic logic. Moreover, written Aristotelian commentaries were sometimes the record of live teaching, in which subjects might be treated weeks apart, and the earlier comments written without knowledge of all the issues that might arise later. On the other hand, masters often repeated the same course of lectures in different years, providing opportunities in the later years for greater coherence and consistency, and we have not only commentaries reporting lectures, but also commentaries revised and prepared for publication, so to speak, by the lecturer or author. I have argued in an earlier paper that William of Ockham, for instance, made a great effort to expound consistent views not only within given scientific works, but within all his works, whether theological, physical, or logical, so that theories developed in theological contexts later shaped what Ockham said on physics and vice versa.[32] I would tend to doubt Edward Grant's claim, then, that the typical fourteenth century Aristotelian commentator did not worry much about consistency from context to context. More research should be done with this question in mind.

Fourteenth-century natural philosophy has often been characterized by its frequent use of arguments *secundum imaginationem*.[33] So, Jean Buridan may ask, what would be the case supposing, *secundum imaginationem*, that God rotated the whole cosmos as one solid body? Would there be any real motion in such a case?[34] The reference to God's power in setting up such hypothetical cases has sometimes been seen as a way of thinking resulting from the condemnations at Paris in 1277, which asserted, in effect, that God can do anything that is not logically impossible.[35] These condemnations have also been linked with the view that fourteenth-century natural philosophers worked in a context shaped by the views of theological faculties. Natural philosophers, it has been asserted, often ascribed only probability or verisimilitude to their theories and not certainty, lest it should turn out that these theories were in conflict with Christian dogma or revelation.[36]

If this was the case, then, science in medieval universities was shaped by the nearness and dominance of theological faculties and concerns. In particular, a lack of assertiveness in science could be ascribed to this factor. I think, however, that this perspective on things has often been exaggerated. It was not only concern about theological oversight that could lead to natural philosophers being tentative or hypothetical. In some cases, merely showing that a case is possible might be sufficient for the argument at hand. So, for example, Nicole Oresme could argue that a given effect need not be magical because it might be explained by the configuration of forms in the body.[37] To make this point, he needed only to propose one way in which this could occur; he need not prove that this is the way it actually happens. In another context, then, he might propose another sort of explanation of the same effect. In doing so, Oresme would not be being inconsistent since his explanations are only possibilities.[38] But neither would his tentativeness spring from theological concerns. In fact, in this context, his major point seems to have been how the theory of the configurations of forms had the power to explain naturally phenomena that might otherwise be supposed to be magical or miraculous.[39]

The problem of theological oversight was not necessarily involved in every case seeming to involve "Averroism" or the so-called doctrine of the double truth. The tendency on the part of some fourteenth-century philosophers to answer questions as philosophers rather than simply as seekers after truth may reflect not concern about theological oversight, but a growing sense that each discipline has its own separate coherence, as the theorems of geometry are coherent with and stand or fall with its definitions, axioms, and postulates. Just as a philosopher might distinguish what he would assert as a philosopher from what he would assert as a Christian, so he might distinguish the perspective of a natural philosopher from that of a mathematician.[40] Within each science it is the scientist's role to assert the propositions that follow from the principles or presuppositions of that science.

Finally, on this point, I think we must temper what we say about Aristotelian commentaries by recognition of the coexistence in time of works like Oresme's On the Configurations of Qualities and Motions or, more importantly, of works like John Dumbleton's Summa of Logic and Natural Philosophy. In a work like Dumbleton's, the topics of natural philosophy could be treated in a logically coherent manner. James Weisheipl has suggested that the first parts of Dumbleton's Summa contain evidence that live teaching lay behind them, while the later parts appear more nearly as purely literary compositions.[41] This is a suggestion that would be worth further investigation.

My belief is, then,—I trust it is not merely wishful thinking—that fourteenth-century Aristotelian commentaries are more carefully considered and consistent than Grant's view implies. This conclusion also has a bearing on the connection of Aristotelian commentaries with arts teaching. Although I should perhaps be chided for suggesting that a work used for undergraduate teaching may contain elements that are purely exercises for students or may contain simplified material that is not the whole truth, whereas a work of research or scholarship would be expected to contain only the most carefully considered and serious claims, nevertheless, if one does make this hypothesis, then the conclusion that Aristotelian commentaries had a wider audience receives further support. In saying this I do not mean to be saying that they were not therefore aimed at the teaching of arts students, but rather that they represent the sort of identity of "teaching" and "research" that I see as characteristic of fourteenth-century universities.

My next and last major question about the genres of medieval scholastic literature on science suggests a caveat to this view, however. The question is, what was the role of works on sophismata and on "calculations" in fourteenth-century science teaching? In addition to hearing lectures on the quadrivium and on works of natural philosophy, the fourteenth-century Oxford undergraduate was required to take part in a certain number of disputations before determining at the time of becoming a bachelor. The statutes of 1268 say that before determining, if the student has responded in the schools, he should respond publically *de sophismatibus* for an entire year before responding *de questione*, and that he should respond *de questione* at least in the summer before the Lent in which he is to determine.[42] The statutes of 1409 say that the determiner should have replied *de questione* at least from the beginning of the Hilary term and that before that time, at least for a year, he should have been a general artist frequenting the *parviso* and disputing and responding.[43] A statute of before 1350 states that the disputations *in parviso* should be logical insofar as possible.[44] In the exercise of determination itself, the student was required to be present in a school during Lent to dispute questions of logic and possibly also of grammar or of other subjects.

What, then, was the role of works like William Heytesbury's *Rules for Solving Sophismata* and of Richard Swineshead's *Book of Calculations*?[45] In addition to logical topics, these works contain theories of the quantification of qualities and of various kinds of motion and therefore seem to represent a kind of natural philosophy or mathematics. Were they part of undergraduate science teaching or should we regard them as the products of the research or scholarship of their authors?

After devoting a lot of attention to this subject, I conclude that these *are* works that should be connected with undergraduate science teaching not in the sense of teaching the undergraduate something about the world, but rather as a part of teaching the students the scholastic methods, especially as this operated in the context of oral argument. One of the essential skills needed by any medieval university student or master was the skill of unraveling arguments to understand their meaning and to check for weaknesses, fallacies, or implications of various sorts. This would apply to law or theology as well as to arts or medicine. In part to teach these skills for oral as well as written use, the student was required to attend and to take part in a number of disputations. Some of these disputations pitted students against each other and involved artificial regulations. For instance, in the so-called "obligations," one student was bound to accept the truth of a given proposition stated at the start—whether it was true in reality or not—and then to accept other statements proposed so long as they were true and not inconsistent with the first proposition, and to do this without falling into self-contradiction.[46] By the mid-fourteenth century it was discovered that use of quantification including infinite series and the like could help to surprise and confuse an opponent in disputation and hence to win the day. When the student used quantification, of course, he was simultaneously practicing mathematical and scientific skills that could serve him in good stead later on. Modern historians of science have noted in this connection advances in the summing of infinite series and in quantification of physics in general.[47] At the same time they have noted the "imaginary" and unrealistic nature of many of the examples considered and have used this evidence to support the characterization of fourteenth-century physics by this *secundum imaginationem* property.

Now I am talking about fairly subtle differences. The same quantitative techniques may appear in works devoted to understanding nature, in works combatting superstition and exhibiting the capacities of mathematics (I am thinking of Oresme), and in works devoted ultimately to exercising students' minds. I think we will understand the calculations and sophismata better, however, if we see that they *are* science for undergraduates, and specifically science for teaching logical thinking and disputation, rather than taking them as the immediate products of scientific concerns and research. Unless we do this we will misunderstand the intentions of the masters who wrote the works, taking as an indication of their views of the world what is rather an indication of their views of how undergraduates should be trained.

Let me conclude then with this somewhat paradoxical result. Most science in medieval universities, I have claimed, may have been science for undergraduates, but it was of such a nature that the concerns of undergraduates, bachelors, and masters of arts could be merged in a single work. Teaching and doing research in natural philosophy in medieval universities were pretty much one and the same. A single genre of literature, the Aristotelian commentary, could play the roles of both the modern textbook and the modern research paper. On the other hand, there were some disputations at Oxford that were intended specifically for undergraduate training, and in connection with these disputations works like Heytesbury's *Rules for Solving Sophismata* were written. If there were, then, any "undergraduate textbooks" in fourteenth-century universities that were not intended for more general use also, presumably these were works like Heytesbury's *Rules* and the later *Libelli sophistarum*,[48] compiled to help prepare students for these disputations. What is paradoxical about this result, and I leave for another occasion discussion of its possible implications, is that modern historians of science and philosophy, including myself, have often found in these "undergraduate textbooks" what they considered some of the newest and most exciting parts of fourteenth-century physics. Does this prove that Heytesbury's *Rules*, for instance, was not solely an undergraduate textbook as I have claimed? Or does it indicate the high quality of fourteenth-century Oxford undergraduate education? Or does it indicate only that some fourteenth-century undergraduates had interests not unlike those of some twentieth-century historians?

NOTES AND REFERENCES

1. Fridericus II, *De arte venandi cum avibus*. Carl Willemsen, ed. (Leipzig, 1942). English translation by C. A. Wood and F. M. Fyfe, *The Art of Falconry* (Palo Alto, CA: Stanford University Press, 1943).
2. See Brian Stock, *Myth and Science in the Twelfth Century. A Study of Bernard Silvester.* (Princeton, NJ: Princeton University Press, 1972); Winthrop Wetherbee, *The Cosmographia of Bernardus Silvestris. A Translation with Introduction and Notes* (New York and London: Columbia University Press, 1973).
3. Stock,[2] p. 13; Wetherbee,[2] p. 20. This need not show a direct connection with Chartres.
4. Stock,[2] p. 13 and note 7; Wetherbee,[2] p. 20
5. Wetherbee,[2] p. 20; Stock[2] (p. 11, note 1) doubts the evidence (a marginal gloss in Oxford, Bodleian MS Laud. Misc. 515, f. 188v) upon which this claim is based.
6. Wetherbee,[2] p. 21. *Cf.* Stock,[2] pp. 273–283, "Conclusion: Literature or Science?"
7. Nancy Siraisi, *Arts and Sciences at Padua. The Studium of Padua before 1350.* Pontifical Institute of Medieval Studies, Texts and Studies, vol. 25 (Toronto, Canada, 1973), pp. 67, 77–94. Siraisi now thinks (as indicated in discussion at the Barnard conference) that she may have exaggerated in this book the importance of astrology in the curriculum at Padua.
8. Strickland Gibson, ed., *Statuta Antiqua Universitatis Oxoniensis* (Oxford: Clarendon Press, 1931), p. 26.
9. *Ibid.*, p. 200.
10. Hastings Rashdall, *The Universities of Europe in the Middle Ages*, 2nd. edit., 3 vols. F. M. Powicke and A. B. Emden, eds. (London: Oxford University Press, 1936), vol. 1, pp. 441–442.
11. Gibson,[8] pp. 32–33:

> vel librum *Poleticorum* vel libros *De animalibus* connumerando libros *De progressu et de motu animalium, Metheoricam*, librosque alios naturales, et *Ethica* Aristotelis audisse complete. . . . Tenetur insuper omnes incepturi, sive prius determinaverint sive non precedente determinacionis actu, si ad incipiendum fuerint presentati, propria iuramento firmare quod sex libros Euclidis, *Arsmetricam* Boycii, *Compotum* cum *Algorismo*, tractatum *De spera* . . . audierint competenter.

Cf. James A. Weisheipl, "Curriculum of the Faculty of Arts at Oxford in the early Fourteenth Century," *Mediaeval Studies* 26 (1964): 161.
12. Gibson,[8] pp. 234–235:

> Arithmetricam per terminum anni, videlicet Boecii; Musicam per terminum anni, videlicet Boecii; Geometriam per duos anni terminos, videlicet librum Geometrie Euclidis, seu Alicen Vitulonemve in perspectivam; Astronomiam per duos terminos anni, videlicet *Theoricam Planetarum* vel Tholomeum in *Almagesti*; Philosophiam Naturalem per tres terminos, videlicet libros *Phisicorum*, vel *Celi et Mundi*, vel *de Proprietatibus Elementorum* aut *Metheororum*, seu *de Vegetabilibus et Plantis*, sive *de Anima*, vel *de Animalibus*, aut aliquem de minutis libris, et hoc de textu Aristotelis. . . .

13. Rashdall,[10] vol. 1, pp. 443–444. In one place the books of mathematics are specified as Sacrobosco's *De sphaera* and one other book.

14. Rashdall,[10] vol. 1, pp. 440–441, says "rhetoric and philosophy are reserved by way of a treat for festivals" and footnotes the text, "non legant in festivis diebus nisi philosophicos et rhetoricas et quadrivialia et *barbarismum* et *ethicam*, si placet, et quartum topichorum." He quotes a statute at Vienna saying, "Quamvis Divinum officium sicut non debemus, ita nolumus perturbare, tamen sanius reputamus quod nostri scholares simul et Baccalarii eciam diebus festiuis visitent Scolas quam Tabernas, dimicent disputando lingua quam gladio, Ergo", etc. *Cf.* Guy Beaujouan, "Motives and Opportunities for Science in Medieval Universities," in *Scientific Change*, Alistair C. Crombie, ed. (New York: Basic Books, 1963), pp. 221–223, who translates the Vienna 1389 statute:

> We think it better for our students to spend their holidays in frequenting the schools rather than the taverns, and to argue with their tongues rather than to fight with their daggers; we are therefore willing that on holidays, after dinner, bachelors of our university should discuss and should read 'gratuitously,' for the love of God, the *computus* and other branches of mathematics, stressing, however, those useful to the service of the Catholic Church.

I particularly recommend this article by Beaujouan for a view of science in medieval universities complementary to that presented here.

15. A. D. Menut and A. J. Denomy, eds., *Nicole Oresme. Le Livre du ciel et du monde* (Madison, Wisconsin: The University of Wisconsin Press, 1968), pp. 3–9.

16. *Ibid.*, p. 731.

17. C. Martin, "Walter Burley," in *Oxford Studies Presented to Daniel Callus* (Oxford: Clarendon Press, 1964), pp. 219–222, 227, 229.

18. *Ibid.*, pp. 219–220.

19. *Ibid.*, p. 221.

20. *Ibid.*, pp. 218–219.

21. John D. North, *Richard of Wallingford. An Edition of his Writings with Introductions, English Translation, and Commentary*, 3 vols. (Oxford: Clarendon Press, 1976), vol. 2, p. 1.

22. *Ibid.*, vol. 2, p. 3.

23. *Ibid.*, vol. 2, p. 1.

24. *Ibid.*, vol. 2, p. 361.

25. *Ibid.*, vol. 2, p. 2.

26. *Ibid.*, vol. 2, p. 15.

27. John E. Murdoch, "The Medieval Euclid: Salient Aspects of the Translations of the *Elements* by Adelard of Bath and Campanus of Novara," in XIIe Congrès International d'Histoire des Sciences, Colloques, in *Revue de synthèse*, 89 (1968): 68–94, esp. 80–83, "The Developed Didacticism of the Medieval Euclid."

28. Stephen K. Victor, "Practical Geometry in the High Middle Ages. Artis Cuiuslibet Consummatio and the Pratike de Geometrie," *Memoirs of the American Philosophical Society* 134 (Philadelphia, 1979): 1–73.

29. Nan L. Hahn, "Medieval Mensuration: Quadrans Vetus and Geometrie Due Sunt Partes Principales . . . ," *Transactions of the American Philosophical Society* 72, part 8 (Philadelphia, 1982): xi–xiv.

30. Edward Grant, "Aristotelianism and the Longevity of the Medieval World View," *History of Science* 16 (1978): 98–100.

31. Ernest Moody, "Galileo and his Precursors," in *Galileo Reappraised*, Carlo Golino, ed. (Berkeley: University of California Press, 1966), pp. 23–43.

32. Edith Dudley Sylla, "Autonomous and Handmaiden Science: St. Thomas Aquinas and William of Ockham on the Physics of the Eucharist," in *The Cultural Context of Medieval Learning*. Proceedings of the First International Colloquium on Philosophy, Science, and Theology in the

Middle Ages, September 1973. John Emery Murdoch and Edith Dudley Sylla, eds. (Dordrecht/Boston: Reidel, 1975). Boston Studies in the Philosophy of Science, **26**: 349–396.

33. See, *e.g.*, Edward Grant, "Late Medieval Thought, Copernicus, and the Scientific Revolution," *Journal of the History of Ideas* **23** (1962): 205 and note 32.

34. John Murdoch and Edith Sylla, "The Science of Motion," in *Science in the Middle Ages*, David C. Lindberg, ed. (Chicago: The University of Chicago Press, 1978), pp. 217–218.

35. See, *e.g.*, Edward Grant, "The Effect of the Condemnation of 1277," in *The Cambridge History of Later Medieval Philosophy*, Norman Kretzmann, Anthony Kenny, and Jan Pinborg, eds. (Cambridge: Cambridge University Press, 1982), pp. 537–539.

36. Edward Grant,[33] pp. 199–201, 206–207.

37. Marshall Clagett, *Nicole Oresme and the Medieval Geometry of Qualities and Motions* (Madison, Wisconsin: University of Wisconsin Press, 1968), pp. 14, 335–387.

38. For some of the ideas expressed in this paragraph, I am grateful to discussions between Edward Grant and Bert Hansen that were part of a session on "That Uncertain Feeling: Medieval Thought About Scientific Truth," which took place at the annual History of Science Society meeting held at Norwalk, Connecticut in October 1983.

39. See Clagett.[37] On page 365, for instance, Oresme says, "Those magicians called necromancers are accustomed to make such fumes. . . . Now one of the causes of such things could be assigned from what was said in the first part of this treatise, namely that such exhalations have this power because of the configuration of the difformity of their qualities."

40. Dumbleton does this in his *Summa Logicae et Philosophiae Naturalis*, when he says, for instance, "things considered in geometry are not admitted unless for the sake of argument or information in other sciences." Cambridge, Peterhouse MS 272, f. 14va.

41. James A. Weisheipl, "Ockham and Some Mertonians," *Medieval Studies* **30** (1968): 202.

42. Gibson,[8] p. 26.

43. *Ibid.*, p. 200.

44. *Ibid.*, p. 27.

45. See Curtis Wilson, *William Heytesbury. Medieval Logic and the Rise of Mathematical Physics* (Madison, Wisconsin: University of Wisconsin Press, 1956); John Murdoch and Edith Sylla, "Swineshead, Richard," in *Dictionary of Scientific Biography*, Charles Coulston Gillispie, ed. (New York: Charles Scribner's Sons, 1976), vol. 13, pp. 184–213.

46. See Eleonore Stump, "Obligations," in *The Cambridge History of Later Medieval Philosophy* (above, note 35), pp. 315–334, and Paul Spade, "Obligations: Developments in the Fourteenth Century," *ibid.*, pp. 335–341.

47. See, *e.g.* Marshall Clagett, *The Science of Mechanics in the Middle Ages* (Madison, Wisconsin: The University of Wisconsin Press, 1959), *passim* (see index).

48. See Edith Sylla, "The Oxford Calculators," in *The Cambridge History of Later Medieval Philosophy* (above, note 35), p. 557.

"Impious Men": Twelfth-Century Attempts to Apply Dialectic to the World of Nature

TINA STIEFEL[a]

The Institute for Research in History
New York, New York 10016

IN THE YEAR 1141 William of St. Thierry angrily wrote, "Let no one think, as certain impious men have, that things contrary to nature — that is, contrary to the accustomed course of nature — cannot occur." And referring to William of Conches' views that we are meant to understand the Biblical account of the creation of woman metaphorically, he went on to say, ". . . As to the creation of woman from the rib of Adam, he [William of Conches] holds the authority of the sacred history in contempt."[1] Such indignation, such passionate attacks resound through time, still being repeated five centuries later by Galileo's enemies. The great metaphysician who wrote that denunciation was responding to the appearance of a totally new mood in Western Europe. A rationalist, critical mode of thinking had sprung into existence. Its "impudent" appearance, as this critic called it, was immediately perceived, understandably, as a threat to the entire fabric of tradition upon which medieval life was based.

My purpose here is to explore a dramatic moment in European history, and in particular the aims and acts of the chief participating characters in the drama as far as we know them. I have previously discussed aspects of this conceptual revolution: specifically, the growth of an idea of the discipline of natural science and the dangers implicit in the critical mentality that accompanied it. My present intention is to focus on the actual reactions and repercussions that the movement awakened and the consequences for the men who initiated it. What, in fact, was this conceptual revolution and what were its implications for medieval society as perceived then?

The most daring of all intellectual enterprises at this time was the concern central to the thinking of a few men, often called cosmologists: William of Conches, Thierry of Chartres, and Adelard of Bath, all of whom wrote in the first half of the twelfth century, and all of whom were concerned with

[a] Mailing address: One Fifth Avenue, New York, NY 10003.

the strict application of critical, analytical thinking to all aspects of natural phenomena, whether astronomy or physiology. Relying on their faith in natural causation and in the atomist structure of the basic substance of the cosmos, they both postulated and attempted to formulate a rational methodology for the investigation of *rerum natura*: they invented for themselves a new discipline — natural science.

For the purpose of this paper, the term rationalism is defined as a pragmatic, unorthodox and nonconformist cast of mind rather than an exercise in abstraction. My term embraces sceptical, empirical, provisional and potentially unsettling forays into every kind of human experience.

The movement described in this study was initiated by men who called themselves *moderni*, who were attempting to find new paths, to break out of the mold of traditional thought. The Greeks had taken the same step seventeen centuries earlier, and it would be another six centuries before the mood reappeared. The rationalist movement I discuss here represents the only appearance of this rare phenomenon between the ancient world and the modern one. It should be emphasized, however, that "rationalist" is used in the loose sense described above, with a wide connotation including the practice of open inquiry and creative hypothesizing, and the impulse to apply them to a broad spectrum of human experience including science, medicine, law, government, theology, literature, and architecture.

My general thesis can be summarized briefly. The prevailing estimate of medieval thinking about natural science is inaccurate on several counts. In my view, productive ideas concerning nature as a fit subject of objective inquiry were articulated in Western Europe before the appearance in the West of Aristotle's scientific corpus in translation. Scientific ideas and approaches, then, were influenced by the third-century Latin version of Plato's *Timaeus* and by scattered bits of Greek science and medicine and Arabic scientific writings. A new approach to the systematic study of science was formulated by the men named and their colleagues a century before Robert Grosseteste and Roger Bacon made their notable scientific contributions. This approach was an adumbration of a scientific methodology — both the mathematical and deductive, and the empirical-inductive techniques, although there was no organized presentation *per se* made of it. (It is the random and scattered nature of the material on science that accounts, I believe, for the fact that its full significance has hitherto been missed.) The twelfth-century thinkers were quite aware of the inferences to be drawn from their work; they promoted their ideas with great courage and energy in the hope of stimulating an intellectual revolution. However, although their works were widely read during the following three centuries, they failed to bring about the revolution they sought, for the times were unpropitious for such radical change.

By the beginning of the twelfth century, this program of studies had so far achieved its aim as to bring about a recognizable increase in the confident expression of a sense of order in human experience. Men experienced a greater sense of dignity than before, as they came to see that by training their intellect they could add to the sum of knowledge and understanding of themselves and the world. As a consequence of the new sense of human dignity expressed at this time, there was a recognition of the dignity and nobility of the created world—of nature, as well. This aspect of twelfth-century humanism naturally followed, for if man sees himself as worthy of dignity, then the natural order of which he is a part must also be noble. An appreciation of the grandeur and beauty of nature becomes itself a human quality. Man takes his place in nature, and human society is seen as part of the grand complex of the natural order which is bound together by rational laws. It began to appear that the whole universe was intelligible and accessible to human reason; nature is now seen as an orderly system, instead of a mysterious, necessarily obscure phenomenon; and man, in understanding the laws of nature, or being potentially capable of so doing, can see himself as the main part, the key-stone of the natural world. It is the possibility of such understanding that gave twelfth-century men the confidence in human powers, a confidence implicit in any humanist movement.[2]

Implicit in this attitude toward mankind is the belief in man's capacity for learning, for development. The twelfth-century cosmologists, like humanists in other periods, hoped for some kind of human progress, and this hope found an outlet in the thirst for scientific knowledge expressed in their work. The concept of progress has not been associated with the twelfth century, but there is clear evidence that something very like it formed a component of the optimistic faith in *ratio* in this period. It began to be apparent that if a man could harness his reasoning faculty and train it to function well, he could with its help learn to understand the world. It was also clear that this faculty was God's gift. Peter Abelard believed that man is like God in having the capacity to reason.[3] Such realizations brought fresh confidence and hope for increasing knowledge of nature and of man, and of the rewards such knowledge could bring. This new optimism was in part due to a sense that the past had been mastered and the future of information about the universe was an open one.

Let us look briefly at the lives of these cosmologists; although little is known about them, there are certain indications which suggest that their biographies have more interest that has been noted. Adelard of Bath, born late in the eleventh century, taught at Paris and Laon. He traveled extensively — perhaps more than any other European writer of his day — in Southern Italy, Sicily, Syria, Palestine, and probably Spain. It is thought that he might have served as an officer of the exchequer at the court of Henry I. He translated many works from Arabic on astronomy, arithmetic, chemistry, and music, as well as practical treatises on such matters as the use of the astrolabe and the art of falconry. There is a story that he performed on the cithara for the queen of France. He was the first to translate Euclid's *Elements* and to introduce the use of Arabic numerals into Europe. His earlier original work *De eodem et diverso* (the title taken from a concept in the *Timaeus*) was probably written *ca.* 1109, and his very popular *Quaestiones naturales*, around 1130. Twenty copies of this document are extant, and it was frequently quoted in subsequent centuries.[4]

William of Conches, of whom Sarton wrote, "[He] has carved for himself a lasting niche in the enduring world of thought and letters by his intellectual courage, originality and thoroughness," was born in Normandy around 1090.[5] We know that he taught at Paris, and it is assumed by most scholars that he was on the faculty of the school of Chartres because John of Salisbury tells of studying with him from 1138 to 1141. This assumption, however, is now under attack.[6] Afterward, in the late 1140s, William was employed as tutor to Henry Plantagenet, the future Henry II of England. His renown as a teacher was very great; he was praised extravagantly by his famous pupil, John of Salisbury.[7]

Excited by the new translations of Greek and Arabic science beginning to appear in the West, William advocated that scientists study these languages so they would be able to read the sources, an idea not taken up again until Grosseteste and Bacon revived it a century later. He was the first to use the noun *elementum* to designate the basic kinds of matter perceptible by the senses. William was cognizant of the new developments in medicine and physiology that were emanating from Toledo and Salerno, and he was well-versed in the scientific writings of Seneca, Pliny, Galen, and Ptolemy. His fame was such that this epitaph was written for him: "Normandy was made famous by his birth, France by his upbringing, Paris by his body, by his mind the world."[8]

Thierry of Chartres, a Breton who taught in Paris and perhaps at Chartres, where he became chancellor in 1141, died around 1155. Little is known about his life, and one fact — that he was the brother of Bernard of Chartres — now appears to be fiction.[9] Some of his students revered this brilliant and controversial master, but many feared his "sharp tongue that cut like a sword."[10] We know that his friends had reason to trust and love him; he defended Peter Abelard at the Council of Soissons in 1121, and he attended the trial of his colleague, the great metaphysician Gilbert de la Porée, at Rheims in 1148, no doubt also in a friendly capacity. We know that Thierry faced criticism and censure for his work. Consider, for example, his despairing and bitter remark that has come down to us: "Fame . . . accuses Thierry everywhere, calls him ignominious names . . . a necromancer and a heretic."[11]

The twelfth-century cosmologists had a firm faith in the progress of knowledge. They envisioned a chain of inquiry into natural science, and they believed that the handing on of scientific study from teacher to student through the generations had a necessarily cumulative effect that would ensure the preservation of scientific knowledge and would provide for unlimited possibilities of advancement. Such confidence was felt about man's ability to go forward in science that, for the first time since antiquity, the belief in the ability of contemporary thinkers to surpass the ancients was clearly and boldly expressed.

The idea of a natural order, with its implied dependence upon the operation of a supreme, rational intelligence, revolutionized medieval thought. "Here reason rules: it makes the universe intelligible; it makes man free . . ."—free to learn, free to develop.[12] Such an orderly view not only altered radically the direction of theological thought in the twelfth century, but provided an incentive for independent scientific work. It also provided a rationale for confidence regarding the power of *ratio* to penetrate the obscurities of nature.

The ability to put what was known of Aristotle's logic to sustained and creative use engendered this belief in the power of the human intellect. Enough of the works on logic were known by the time of Erigena to affect his thought, but the full force of the intellectual revolution that the discovery of logic engendered began later. The writings on logic which Boethius had assembled were already so thoroughly assimilated by the last quarter of the eleventh century, having been enthusiastically taught at the turn of the century by Gerbert (d. 1003), that the notion of applying dialectic to all kinds of knowledge influenced by the dialectitions of the period. Under Berengar of Tours (d. 1088), this practice emerged as a conscious technique of rational thought.

> He is ready to state positively that reason, and not authority, is mistress and judge. Dialectic is the art of arts, and it is the sign of an eminent mind that it turns in all things to dialectic. Anyone who does not do so abandons his principle glory, for it is by reason that man resembles God. He intends, therefore, to have resort to dialectic in all things, because dialectic is the exercise of reason, and reason is incomparably superior to authority when it is a question of ascertaining the truth.[13]

Berengar's pioneering effort to subject Christian tenets to rational scrutiny were recognized and opposed by Lanfranc, who accused him of wanting to approach all problems through reason and forsake tradition.

This passion for applying rational techniques to every kind of thought had a direct impact on science when Roscelin of Compiègne, a follower of the popular art of dialectic at the turn of the century, enunciated the nominalist theory, which includes the idea that objects of our senses have a singular, individual identity or existence. He said that in nature the individual alone exists. Roscelin's position clearly implies the value of concrete, material observation of the *res*, the phenomena under consideration.

Anselm of Aosta utilized rationalism in his theological speculations; his "proofs" of God's existence are evidence of the power of this intellectual movement, for it is at this time that the habit of seeking proof for statements requiring belief was established. Anselm held that the teachings of faith were ultimately rational, and he expressed a firm trust in the importance to the Christian of the functioning intellect. In *Cur Deus Homo* (c. 2), he wrote, "It seems to me to be negligent if, being strong believers, we are not zealous to understand our beliefs."[14]

From Berengar on through Roscelin and Anselm, the new intense interest in *ratio*, in the application of the rules of dialectic, was focused with increasing sharpness upon philosophical problems related to Christian doctrine. Hence it is the more remarkable that the cosmologists were, for the most part, able to leave this use of Aristotelian logic alone, and to invent instead a radically different use for the novel technique. It is precisely in this creative act that the importance of the cosmologists in the history of ideas lies. They reasoned, in effect, thus: God has given us His gift of the capacity to reason. When we develop this divine gift, we can use it productively and, in doing so, enhance what is most distinctively human in us. The created universe operates according to basically rational principles, having been deliberately so designed by its creator. Man, by virtue of his capacity for reasoning, and also because he is a part of a rational nature, is fitted to understand the ways in which the natural world functions. The cosmologists arrived at this conclusion, which implied the need for a new discipline, natural science, by means of a perfect Aristotelian syllogism.[15]

Adelard of Bath often expressed his belief in man's innate capacity for rational thought. At one point he presented an almost Darwinian argument[16]; at another point Adelard unconsciously echoed Aristotle: "Man is a rational animal and for that reason he is sociable as well; and he is innately fitted thereby for the two operations of deliberation and action." Adelard thought that man's rational faculty helped him to deal with his emotions: "Reason was a necessary function of the soul as an aid for moderation, of controlling the passions, and was not lacking from it."[17] Again he wrote, "In war, anger provides the motivation. In peace, reason does the pacifying."[18] William of Conches saw the reasoning mechanism as underlying all mental functions.[19] Since nature was seen as amenable to predictability in terms of cause and effect, it was clearly akin to human rationality.[20]

It is not easy for the twentieth-century historian to apprehend the force of this revolution in thought. No longer was man seen as a puppet of mysterious forces referable only to the divine will: man and nature, by God's plan, were inextricably united in an ultimately intelligible universe.

Hugh of St. Victor testified to the rationality of the creator's plan and man's relation to it:

> It seems to us — though we do not wish to make any rash definition on the point — that it takes away nothing from the creator's omnipotence if one says that he brought his work to completion across intervals of time; . . . in all things God proposed doing, he must have kept especially to that mode of action which best served the need and convenience of his rational creature.[21]

Honorius of Autun was in agreement with this view and went on to develop the idea of man as an intrinsic part of nature:

> Whence came the corporeal substance used in man's creation? From the four elements, and for this reason, man is called a microcosm — that is, a lesser world; for from earth he has his flesh, from water his blood, from air his breath and from fire his warmth.[22]

This twelfth-century revival of the *Timaeus'* idea of man as microcosm was made more complex later in the century by Alan of Lille. He spoke of man as a kind of lesser god, capable of creative work and having, in this capacity as *homo artifex*, a function closely related to operant nature, *opus naturae*, as well as to God as shaping agent, *opus creatoris*.

> His [God's] working is one, whereas mine is many; his work stands of itself, whereas mine fails from within; his work is a thing of wonder, mine mutable. His is without origin, whereas I was brought forth; he is a maker, I made; he is the workman behind my work, and I am the work of that workman; he makes his work out of nothing, I beg mine from some source; he works in his own name, and I in his. . . . And, in order that you may recognize that my power is powerless in contrast to the divine power, know that my effort is defective and my energy worthless.[23]

In spite of its tone of genuine Christian humility, this passage expresses considerable confidence in man's rational capacities, a confidence that did not exist before the twelfth century — not, in fact, since the ancient world.

The new attention accorded to the status of man within the natural realm prompted attempts to explain why it was that our species was created. In comparing man as maker with the creator, Alan was carrying out the macro-cosm/microcosm theme of the *Timaeus* and developing it almost to the point of *hubris*. Thierry of Chartres expressed the same idea at the close of his commentary on Genesis, when he wrote that his book had unity and that this unity was a correlative of the unity of creation. (Just such a consciously constructed unity is seen in Dante's *Commedia*, in which the poet's function is conceived in accordance with the root meaning of the word "poet": ποιητής, one who makes or creates.)

That man is part of a rationally planned and executed universe was a liberating notion for the twelfth-century writers. It is not surprising that, with the freshly perceived relevance of this idea to the new sensitivity to nature at this time, the concept of a reified nature — the hypostatizing of the natural world as "*natura*"—was also vigorously revived. The twelfth century was familiar with this classical conception.[24]

The poet Bernard Sylvestris employed *natura* in the delineation of his vision of a Neoplatonist cosmology. In his poem *De mundi universitate, Natura* is seen as shaping matter (*silva*); she implants the earth with the seeds of life; she rules generation throughout the range of living beings; she is "*mater generationis*," "The most ancient aspect of nature, the inexhaustible womb of generation." This work was very popular in its time, as the number of extant manuscripts attests. The goddess *Natura* plays a central role in this curious poem, much of which is of pagan provenance, and this centrality, combined with the evidence of the work's interest for its contemporary readers, reflects the importance of the concept of nature in this period.[25]

Although twelfth-century men were cognizant of the kinds of objections being made to the practice of scientific inquiry, the first task that they approached was to imbue their contemporaries with their own trust in a rational nature. In his commentary on Genesis, Thierry of Chartres explained that the created world was so constructed, because it was created by God alone, in order to exhibit in the highest degree a rational and beautiful order. Thierry believed that this beauty and order (*rationabiliter*) existed as a consequence of being in accordance with the wisdom of its creator. When Adelard of Bath opposed the common opinion that the fact that grass exists can be explained only as "a wonderful effect of the wonderful divine will," he said that of course it was the creator's will that grass exists, "but it is also not without a natural reason"; and he reiterated, "but there is nothing [in nature] without *ratio*."[26]

Hugh of St. Victor wrote with certainty of the rational principle underlying the universe:

The ordered disposition of things from top to bottom in the network of this

universe . . . is so arranged that, among all the things that exist [in nature] nothing
is unconnected or separable by nature, or external.[27]

Thierry of Chartres said, "The world would seem to have causes for its existence, and also to have come into existence in a predictable sequence in time.
This existence and this order can be shown to be rational."[28]

When questioned about the nature of the stars, Adelard of Bath replied,
"For whatever is in or on these [stars], I consider that they are the product
of a rational nature."[29] Seizing on a Biblical reference to drive his point home,
he wrote:

> . . . whence it happens that the visible universe is subject to quantification and
> measure and is so by necessity. For each thing exists [in nature] either as one
> or as many. The immense universe itself is also defined by the fact of its being
> limited by its very nature.[30]

Measurability defines the universe by means of a terminal boundary; the universe is hence finite. The emphasis is on nature as understandable by its property of being limited; this has two implications. It can be understood because
it has limits; and, because quantification and measure are properties of all
natural phenomena, mathematics is an important tool for attaining knowledge of nature. The use of the adjective "immense" reflects the confidence
Adelard felt in his scientific enterprise, as if he were saying that the cosmos
may seem dauntingly large, but is, in fact, open to human comprehension
by its very construction, so that there is no need to feel helpless before it.

Adelard believed that rational man ought to be putting his divine gift of
reason to use, and also that in studying nature he was proving his right and
exercising his privilege of being a part of it.[31] He felt a passionate enthusiasm
for science, and was sensitive to the esthetic appeal of nature.

Elsewhere, Adelard shows his deep certainty of the essential rightness and
esthetic fitness of everything in nature. When queried about the comeliness
of a face disfigured by the effects of a head cold, he replied, "There is nothing
in nature either dirty or unsightly. But, whatever is contrary to the *ratio* of
nature, however much is added to the handsomeness of a face by painting,
we have the right to call dirty and ugly."[32] His dislike of artifice, together with
this strong championing of the rightness of nature, is here rather like Rousseau's. There is no attempt to idealize nature, but only a desire to hold up
for admiration the natural world, the cosmos: an orderly, functioning entity,
pleasing to the eye and to the imagination.[33]

The new awareness of the rationality of the universe that the twelfth century absorbed from the *Timaeus* brought with it a feeling of awe and admira-

tion for a system so beautifully designed. The cosmologists rejoiced in this feeling, and were thereby released from the attitude of *contemptus mundi*. Richard of St. Victor thought that the celebration of the *ornatus* of the created world was a necessary stage for Victorines in their ascent to an understanding of God.[34]

To sum up, the twelfth-century scientists, or cosmologists, urged the scientific investigation of the natural world as the duty of a Christian, undertaken in gratitude for the magnificent gift of the cosmos. Such investigation was a suitable response of rational man, aware of his kinship with nature and with his God in terms of his rational capacity. This use of reason would be a source of pleasure and happiness, and a means of celebrating the glory of God's work. Their experience was channeled into religious expression; for the first time since antiquity — and more than a century before Roger Bacon articulated the thought — these writers made explicit the idea that the ultimate purpose of examining and investigating the created world was not primarily for the acquisition of scientific knowledge for its own sake, but was to help men reach a higher level of understanding of the creator.

Although we use the term "program" for the working out of the notion of natural science as a separate branch of study, it should not be inferred that the twelfth-century men who were concerned with this matter were organized or systematic in their exposition of their program. For though the cardinal points of a viable approach to science as a discipline were understood to the extent that was conceivable before the nineteenth century, the necessity for systematizing was not seen then; and so we are compelled to impose a kind of order upon what was, for the most part, random and unconnected, dispersed throughout these men's works. Even if the cosmologists did not consciously set about framing a scientific program, they did have a coherent plan of procedure, which justifies our use of the word "program." Their idea included a community of *physici* who worked cooperatively to solve scientific problems.

The value of communal efforts for the practice of science was well understood. If the cosmologists could not predict the teamwork of the modern laboratory, they at least saw the importance of sharing their investigations and approaching scientific problems in the company of their colleagues. Mutual discussion and reflection was most helpful to the inquiring scientist. Adelard compliments his nephew's theorizing on a thorny question, "Thus reason advances and we think alike." In a discussion with a colleague on the importance of logically based theories in natural science, he said, "Continue, therefore between you and me, reason only shall be judge . . . since you are proceeding according to the rational method, I will give reason and take it too." He encouraged his colleague, ". . . And since the first question has been solved by reason, if you have doubts about others, speak up."[35] Exchanging theories and hypotheses helps to correct errors, curbs a too inventive imagination and pools the intellectual resources of the group, while encouraging bold, independent thinking.

From the time of Berengar the possible application of the art of dialectic to the study of theology was considered. After Anselm, the movement to rationalize theology gained momentum, and as it became common for men to attempt to order every kind of knowledge as a system of deductions like mathematics, starting from undemonstrable first principles and proceeding to the proposition to be demonstrated, some began to apply this method to theology. A need was felt by the cosmologists and their followers to come to terms with the question of the relation between the study of natural science and the study of theology. One way to reconcile them was to regard reason as a valuable aid in tracing the craftmanship of the creator and to see its highest function as seeking knowledge of the causes of things in the service of God.

The scientists of the early twelfth century offered tentative suggestions regarding the occurrence of miracles in a universe tightly governed by rational laws. William of Conches saw that "the world is an ordered aggregation of creatures."[36] He tried to clarify the rational principles underlying nature and at the same time account for the possibility of God's acting outside it.[37] Peter Abelard also tried to deal with points at which nature's powers and God's diverge:

> Perhaps someone will ask too by what power of nature this came to be. First, I will reply that when we require to assign the power of nature or natural causes to certain effects of things, we by no means do so in a manner resembling God's first operation in constituting the world, when only the will of God had the force of nature in creating things. . . . We go on to examine the power of nature . . . so that the constitution or development of everything that originates without miracles can be adequately accounted for.[38]

William of Conches was at some pains to reserve for God all due praise for the creation of man while giving to nature the credit for effecting the generation of men.[39]

Although the critical attitude toward authorities cultivated by the cosmologists did not extend to Scripture, it was inevitable that these men would find themselves extending their new mental habit of critical analysis to those portions of Genesis which offered an account of the creation of the world. Thierry of Chartres introduced his commentary on the first chapter of this book, "This is an exegetical study of the first portion of Genesis from the point of view of an investigator of natural processes (*secundum physicam*) and of the literal meaning of the text."[40] This simple statement of intention was of historic importance, for it was the first conscious and deliberate attempt to analyze a part of the Bible as a scientist, a man in search of rational explanations. We must keep in mind that the Bible was the only account of the origins of the universe known to the twelfth century except for that given in the *Timaeus*.

Resolutely, Thierry commenced his study, "In the beginning God created the heaven and the earth," and then proceeded in the way he had set himself to go:

> . . . As if He might say: He first created heaven and earth. For He wanted another to understand through this statement, that when He said, "In the beginning He created," it is to be understood that He created nothing prior to these and that this act was the creation of these two entities simultaneously. But I will attempt to demonstrate both what is meant by "heaven" and "earth," and also show the manner of this coming into being of these natural phenomena in accordance with reason.[41]

Thierry went on to explain that Moses' purpose in writing this book was to show how God alone accomplished His great task and so carried it out as to illustrate perfectly His own wisdom. This universe, therefore, is necessarily so constructed as to exhibit in the highest degree a rational and beautiful order. This beauty and this logic (*rationabiliter*) exist in consequence of God's paramount reason. When Moses, for example, says "In the beginning . . ." what he literally means is that the first thing that God created was actually matter — the basic substratum of things that includes four elements: fire, air, earth and water. Afterwards, by a natural process, matter took on distinguishable forms and in these forms the four elements were put in motion.

From this account of a portion of Thierry's work, it is possible to discern his mode of operation. Bit by bit he works his way through the sacred text, explicating each statement *secundum physicam rationes* with highly original and ingenious interpretations which confidently offer a logical explanation of each point. Thierry presented the concept of a chain of causality in the entire process of creation from primordial matter to its most complex form, man; his purpose was to bring out the logical and, so to speak, automatic nature of the creation of the world. There is something resembling the notion of a chain reaction which got under way after God performed the initial act of creating primal matter. "Reason" continues to produce in nature the whole, complex cosmos that God envisioned before he acted. In the combining and mutual reacting of the elements of basic matter, Thierry saw the mechanism of a continual creation occurring in time — a theory similar to current theories on the chemical beginnings of life. He postulated transformations of matter in the direction of increasing differentiation, from a thoroughly mixed state to clearly discrete entities.

William of Conches also took on portions of the Bible for scrutinizing in a scientific manner. His comments are as cogent as Thierry's, but his manner is less organized; he had no wish to cover everything on nature systematically. When he was unable to make the Biblical statement yield a reasonable explanation, he said so. He wrote, ". . . The Divine Scriptures tell us that God created man from the dirt of the earth; but this must not be believed — that the mind which is quasi-divine, light and elegant was made from dirt."[42] The view of William of Conches that the creation of Eve was effected through the action of natural forces on matter opened him to the charge of naturalism, the theory that nature works independently of God. William avoided the charge by saying that nature's capacity to function was a divine gift. He thought that one ought not to invoke God's omnipotence in giving scientific explanations.

Medieval men inherited from Augustine a Christian tradition that encouraged a consideration of the nature of things, and consequently Western Christendom was predisposed to value the natural world as sacramental and symbolic of Christian truths. By the twelfth century, however, some men had begun to realize that the study of natural causes had a legitimate interest of its own, quite apart from nature's emblematic function. Adelard of Bath was aware that the scientific approach must differ from the theological, that natural philosophy must invent a different set of concepts to express an important distinction of thought.

William of Conches was eager to defend his literal examination of statements in Genesis, and he devised an argument that advanced a distinction between God's power to create, or do anything He wills, and the extent of natural forces. "But someone will say, 'Isn't it the Creator's work that man is born from man?' to which I reply, 'I am not taking anything away from God.' "[43] To draw a line between the two kinds and degrees of power was not easy, and many failed to comprehend it; William lost patience with those men and hotly chastized them for ignorance and arrogance. Neither William nor Thierry was able to convince his critics and the meager evidence we have suggests that the critics gained the upper hand and eventually brought them to their knees. William wrote derisively of the enemies of the new science; Peter Abelard commented, "Whatever they do not understand they call foolishness; whatever they cannot grasp they judge to be insanity."[44] William said that it was not the purpose of the Bible to teach us about nature, but the purpose of philosophy. Again, he asked, "How are we contrary to the Divine Scripture if, concerning that which it states to have been done, we explain the manner in which it was done?"[45] Peter Abelard supported William of Conches' argument that science added to, rather than subtracted from, divine power.

The problem of establishing a division between divine power and nature's forces was peculiarly relevant to medieval concerns because it involved the validation of the occurrence of miracles. We saw how the twelfth-century scientists at times clearly delimited the study of nature from the larger and more central subject, theology. The earliest of the cosmologists, Adelard of Bath, felt sufficient confidence in nature's basic rationality to say,

> Wherefore, since a wise God is unwilling to abolish the orderly workings of nature, it is not even possible for such a change of position [relating to the idea that nature is not ordered] to occur, and among philosophers it is agreed that any upsetting of this spirit of order is least likely to occur.[46]

An occasional diatribe against the new science can still be found to give us an idea of what the cosmologists were up against. "Let no one impiously think, as certain impious men have, that things contrary to nature — that is, contrary to the accustomed course of nature — cannot occur," wrote William of St. Thierry.[47] Absolom of St. Victor roundly condemned this prying into the "composition of the globe, the nature of the elements, the location of the stars, the nature of animals, the violence of the wind, the life-processes of plants and of roots."[48] The metaphysician William of St. Thierry was outspoken in his total rejection of, and deep hostility toward the work of the men promoting the new science:

> As to the creation of woman from the rib of Adam, [William of Conches] holds the authority of the sacred history in contempt . . .; by interpreting that history from the point of view of physical science, he arrogantly prefers the ideas he invents to the truth that the history contains, and in so doing he makes light of a great mystery. . . . Moreover, after the theology of Peter Abelard, William of Conches produced a new philosophy, confirming and strengthening whatever the former had said; and with greater impudence adding to it much of his own which Abelard has not said, William's worthless innovations.[49]

Some degree of persecution was experienced by Thierry of Chartres as well as by William of Conches and Peter Abelard; long after Thierry had been forced out of his teaching position, he wrote bitterly, *"Ecce Theodoricus Brito — homo barbaricae nationis, verbi insulus, corpore et mente incompositus, mendacem de te se vocat. . . ."*[50]

Certainly Peter Abelard's life needs no comment here — and it is surprising to discover that his tragic end in some degree was shared by the men we have been talking about. At his death in 1142, his admirers still existed, as the testimony of the epitaph written by Peter the Venerable, Abbot of Cluny, shows:

The Socrates of the French, the great Plato of the West, our own Aristotle, either the equal or the better of any logician who ever lived; recognized in the world as the prince of the erudite; versatile in his intellect, subtle and sharp, conquering all through the force of his reason and the art of his speaking . . .[51]

The wild enthusiasm expressed here gives us a hint of the excitement generated by the personality and achievement of Peter Abelard. It also, I think, reflects the warm response which the cosmologists elicited from their followers. The twelfth-century rationalists deserve a wider appreciation, a more accurate assessment, and a more sympathetic treatment than they have received, either in their own time or in ours.

NOTES AND REFERENCES

1. "De erroribus Gulielmi de Conchis," J.P. Migne, *Patrologia Latina* (hereafter cited as *PL*) vol. 180, col. 339. All translations are mine unless otherwise noted.

2. I have written the following articles on aspects of this problem: "Science, Reason and Faith in the Twelfth Century: The Cosmologists' Attack on Tradition," *Journal of European Studies* (1965)vi:1–16. "The Heresy of Science: A Twelfth Century Conceptual Revolution," *Isis* (1977) 68:347–362. "Twelfth Century Matter for Metaphor: The Material View of Plato's Timaeus" *British Journal for The History of Science* (1984), forthcoming.

3. Peter Abelard, *Theologia christiana*, *PL* vol. 178, col. 1315.

4. In the thirteenth century Roger Bacon referred to it; and as late as the fifteenth century it was quoted by Pico della Mirandola. For further details on Adelard's work, see Charles H. Haskins, *Studies in the History of Mediaeval Science*, 2nd edit. (Cambridge, MA: Harvard University Press, 1927), pp. 20–42.

5. George Sarton, *Introduction to the History of Science*, vol. II. (Baltimore: Williams & Wilkins Co., 1931), pp. 71ff.

6. For a comprehensive discussion of this question see R.W. Southern, "The Schools of Paris and of Chartres," in *Renaissance and Renewal in the Twelfth Century*. Robert L. Benson and Giles Constable, eds. (Cambridge, MA: Harvard University Press, 1982), pp. 129 ff.

7. See John of Salisbury, *Metalogicon*, 1.5 and 2.10. It is now believed that William of Conches probably knew most of the philosophers of his day. On this point, see *Guillaume de Conches: Glosae in Juvenalem*, Bradford Wilson, ed. (Paris: Vrin, 1980) p. 83.

8. "Eius praeclaret natu Normannia, victu Francia, Parisius corpore, mente polus." Jules Alexandre Clerval, *Les écoles de Chartres au moyen-âge* (Paris: A. Picard et fils, 1895) p. 182.

9. See R.W. Southern, *Medieval Humanism* (New York: Harper & Row Publishers, 1970), pp. 68ff.

10. "Metamorphosis Goliae" in Thomas Wright, *The Latin Poems Commonly Attributed to Walter Mapes* (London: D. B. Nichols and Son, 1841), p. 28.

11. Nikolaus M. Häring, "Thierry of Chartres and Dominicus Gundissalinus" in *Medieval Studies* 26 (1964): 278.

12. Southern,[9] p. 58.

13. Berengarius, *De sacra coena adversus Lanfrancum*, A.F. and F.Th. Vischer, eds. (Berlin: S. J. Josephy, 1834). (Passage translated by David Knowles in *The Evolution of Medieval Thought*. [New York: Vintage Press, 1964, p. 95].)

14. "Neglegentia mihi videtur, si postquam confirmati sumus in fide, non studemus, quod credidimus intelligere."

15. Winthrop Wetherbee says:

The development of a theology devoted to seeking the truth of scripture through the employment of the arts and the study of nature stimulated serious debate over the extent to which dialectic and secular learning might be allowed to encroach upon the traditional province of exegesis . . .

Platonism and Poetry in the Twelfth Century: The Literary Influences of the School of Chartres (Princeton, NJ: Princeton University Press, 1972), p. 17.

16. Although man is not armed by nature nor is naturally swiftest in flight, yet he has that which is better by far and worth more — that is, reason. For by the possession of this function he exceeds the beasts to such a degree that he subdues them. . . . You see therefore, how much the gift of reason surpasses mere physical equipment.

Adelard of Bath, *Quaestiones Naturales*. M. Müller. ed.. *Beiträge zur Geschichte der Philosophie und Theologie des Mittelalters*, vol. 31.2 (Münster: Aschendorff, 1934) p. 20.

17. Adelard of Bath, *De eodem et diverso*, H. Willner, ed. *Beiträge zur Geschichte der Philosophie und Theologie des Mittelalters*, vol. 4.1 (Münster: Aschendorf, 1903).

18. Adelard of Bath, *Quaes.*: 21.[16]

19. *De philosophia mundi* (under Honorius Augustod), *PL* vol. 90, cols. 1127–1178, ch. XXXIV.

20. Father Chenu wrote that the twelfth-century men experienced

. . . the realization [of the kinship between nature and man] . . . when they thought of themselves as confronting an external, present, intelligible and active reality as they might confront a partner . . . whose might and whose decrees called for accommodation or conflict — a realization which struck them at the very moment when, with no less a shock, they reflected that they were themselves also bits of the cosmos they were ready to master.

M.D. Chenu, *Nature, Man and Society*, trans. by J. Taylor and L. Little (Chicago: University of Chicago Press, 1968), p. 5.

21. Hugh of St. Victor, *De sacramentis Christianae fidei*, *PL* vol. 176, col. 188.

22. Honorius of Autun, *Elucidarium*, *PL* vol. 172, col. 116.

23. Alan of Lille, *De Planctu Naturae*, *PL* vol. 210, col. 445.

24. Ovid used it in the *Metamorphoses* (1, 1-30); he began his cosmogony by a description of Chaos and mentioned "a god or milder nature" bringing an end to the conflict that Chaos is. In Lucretius (*De rerum natura* 2, 603) Venus is the creator of universal life, and she governs the nature of things; he calls her *"naturae creatrix."* Martianus Capella (*De nuptiis phil.* 1, 18) speaks of *Natura* as *"generationum omnium mater."* This universal goddess was at once the symbol of an intellectual concept, and a divinity evoking a genuine feeling of reverence in the fifth century.

25. In the twelfth century *Natura* appeared frequently in many kinds of writings. For example, John of Salisbury wrote, *"Unica causarum ratio divina voluntas, quam Plato naturae nomine saepe vocat. Illius imperio servit natura creata, ordoque causarum totus adhaeret ei."* John of Salisbury, *Entheticus*, *PL* vol. 199, col. 978.

26. *Quaes.*: 6.[16]

27. *De sacra.*, col. 206.[21]

28. *De septem diebus et sex operibus*, M. Hauréau, ed., *Notices et Extraits des Manuscrits de la Bibliothèque Nationale*, vol. 32, part 2, 167-186, Paris (1888): 172.

29. *Quaes.*: 63.[16]

30. *Eodem*: 23.[17]

31. You say that if anyone living in a house is ignorant of what it is made, of its size and quality and layout, he is unworthy of its shelter: and that, just so, if anyone born and educated in the residence of this world neglects learning the plan underlying its

marvellous beauty, upon attaining the age of discretion, he is unworthy, and, were it possible, deserves to be cast out of it.

Astrolabium, praef., MS Cambridge, McClean 165, f. 81.

32. *Quaes.*: 23.[16]

33. Another expression of wonder at the beauty of the universe as a work of great skill is by a coeval of the cosmologists, the monk known as Pseudo-Hugh of St. Victor; he writes of ". . . the beauty of the universe, the intricate contrivance of heaven and earth, this marvellous and delightful work. . . ." in *Liber de stabilitate animae*, PL vol. 213, col. 917. There is in all these exclamations a sense of the complicated and elaborately interconnected nature of this creation; the cosmologists experienced its enormous complexity and beauty as twin aspects.

34. *Benjamin Maior*, PL vol. 196, col. 70, ch. 1.6.

35. *Quaes.*: 6 ff.[16]

36. *Glosae super Platonem*, E. Jeauneau, ed. in *Textes Philosophiques du Moyen Âge*, vol. XIII (Paris, 1965), p. 125.

37. In his commentary on Plato's *Timaeus* he wrote:

> Having shown that nothing exists without a cause, Plato now narrows the discussion to the derivation of effect from efficient cause. It must be realized that every work is the work of the *Creator*, or of *Nature*, or the work of a human artisan imitating nature. The work of the Creator is the first creation without pre-existing material, for example the creation of the elements or spirits, or it is the things we see happen contrary to the accustomed course of nature, as the Virgin Birth and the like.

Glosae: 104.[36]

38. *Exposito in Hexaemeron*, PL vol. 178, col. 746.

39. Someone will allege that this [the idea of nature's being responsible for the generation of men] is to derogate from the divine power. To such we shall reply that, on the contrary, it adds to the divine power because to that power we attribute not only the giving of a productive nature of things but the creation of the human body through the operation of such nature.

Phil. mun. Lib. 1, XXIII.[19]

40. *De septem diebus*: 172.[28]

41. *De septem diebus*: 177.[28]

42. *Phil. mun.* Lib. 1, XXXIII.[19]

43. *Phil. mun.* Lib. 1, XXIII.[19]

44. *Epistulae XIII*, PL vol. 178, col. 353A.

45. *Phil. mun.* Lib. 1, XXXIII.[19]

46. *Quaes.*: 66.[16]

47. *De erroribus.*[1]

48. *De erroribus*,[1] col. 339.

49. *De erroribus*,[1] cols. 339–340.

50. Häring,[11] "Thierry and Dominicus," p. 277.

51. Gallorum Socrates, Plato maximus Hesperiarum, noster Aristoteles, logicis quicunque fuerunt aut par aut melior; studiorum cognitus orbi princeps, ingenio varius, subtilis et acer, omnia vi superans rationis et arte loquendi, Abaelardus erat.

PL vol. 178, col. 103.

Optics through the Eyes of the Medieval Churchmen[a]

SAMUEL DEVONS

Barnard-Columbia History of Physics Laboratory
Columbia University
New York, New York 10027

> *Only a misguided mind tries to introduce religion into*
> *science; more misguided still is he who attempts*
> *to introduce science into religion.*
> — LOUIS PASTEUR, *ca.* 1880

B Y THE LATE NINETEENTH CENTURY the separation of science and religion had become, *de facto*, accepted, by both parties: Pasteur's stern admonition epitomized the views of many who were publicly prominent in science, and in private firmly, perhaps even devoutly religious. It expresses a tacitly accepted attitude to an issue that had already then and certainly has today ceased to arouse intellectual excitement or spiritual passion. It is hard to conceive of the actual articulation of science in an all-pervasive spiritual-religious context. How differently matters stood some seven or eight hundred years ago, at the time when medieval scholars and churchmen began to display and develop their newly aroused interest in the workings of the natural world! It was not just that the medieval church was so potent and ubiquitous an institution that no spiritual, cultural, or intellectual activity could fail to come under its influence, whether to be fostered under its wing, guided by its doctrine and teaching, controlled by its authority or suppressed by its power; but also that the pioneers, the advocates and the expositors of the new scientific activity, and of optical science especially, were themselves leaders of the church or of the monastic and mendicant orders. Science and religion, faith and reason, the secular and the sacred were united, even harmonized in one and the same individual, and in a single ultimate goal — the perfection of the soul in accord with Divine will and command.

From its rudimentary beginnings in the twelfth century, to its fullest flowering at the end of the thirteenth, the optical science of Western Christendom exhibits rapid change and development, but throughout, its close

[a] This talk was supplemented by demonstrations in the History of Physics Laboratory. The work was assisted by grant (SES 8308480) from the National Science Foundation.

205

affiliation to the Church and its service to the faith of churchmen persists. In the late twelth century when Alexander Neckam,[1] Augustinian Canon and Abbot of Cirencester was expounding what was little beyond casual observation blended with hearsay and embellished with spiritual allegory, the great influx of translations of Islamic writings, which so vastly enlarged the medieval awareness of Greek, Alexandrian, and Arabic science, had scarce begun. The science of optics was a paltry thing wholly incommensurate with the imposing body of Christian doctrine, with which it could be associated only in a very minor subordinate capacity.

Some thirty years later, *ca.* 1230, when Robert Grosseteste,[2] the leading Franciscan at Oxford, and later Bishop of Lincoln (one of the most important Sees in England), was formulating the pattern and purpose of the new science, a major part of the Islamic-Greek legacy had reached the west. This new science, and optics in particular (although but one of Grosseteste's concerns, which ranged from criticism of the papacy and ecclesiatical reform to the organization of philosophical, metaphysical, and theological studies at Oxford), was a substantial body of learning, heir to the scholarship of more than a thousand years, with a stature which, if not challenging, was at least no longer dwarfed by the imposing body of doctrine.

Grosseteste's influence was a seminal one for thirteenth century scholarship. His most illustrious disciple, Roger Bacon,[3] the Franciscan "Dr. Mirabilis" of Oxford and Paris, pursued the new science, proclaiming its spiritual values with a fervor unmatched by his contemporaries; he was sustained no doubt by a direct commission from no less than the Pope himself, Clement IV. The rival Dominicans, or some at least, seem at one with their Franciscan brothers in their eagerness to assimilate, elaborate, and disseminate the new learning. The renowned Albertus Magnus,[4] "Dr. Universalis" of Cologne, Provincial of the German Dominican Order and Bishop of Ratisbon, took as his task no less than to make available and clear for the whole Christian world the entire Aristotelian corpus; and to examine not only its compatibility but even its harmony with Christian doctrine. Albertus' disciple, Witelo,[5] a Silesian Dominican at the Papal court at Viterbo assimilated, elaborated and expounded for his Latin contemporaries the bulk of the Greek-Islamic and contemporary medieval knowledge of optics in a work that was to be influential in the west for several centuries. Less technical, more popular, and probably more influential in spreading knowledge of the new optical science was the compilation of no less a dignitary than John Pecham,[6] Archbishop of Canterbury. The last of the scientist-brothers, and the one who at the turn of the century brought medieval optics to its highest point was one Theodoric of Freiberg,[7] Dominican leader of German Preachers. His crowning achievement was the resolution of the Rainbow Mystery, a problem that intrigued and challenged philosophers, on and off, since the time of Aristotle and had become a sort of capstone of any optical treatise.

In this unique thirteenth century association, approaching a partnership of Religion and Science, there was never any question of where final authority lay. Nor had this union been achieved without struggle and challenge. Already two centuries earlier, when Peter Abelard was urging that faith should at least be tested for consistency with reason — not that true faith should or could in any way be subordinate to reason, but that reason might be used to clarify or cleanse the spiritual from unworthy contamination — the suspected challenge to doctrinal faith was branded by the churchmen as blasphemous impudence. The faithful could reply in the words of the revered Augustine, echoed by St. Anselm: "I do not seek to understand in order to believe; I believe in order to understand." In 1272 at the very time that medieval science was in full flower, at the University of Paris, the center of scholastic learning, it appeared necessary to issue a Statute addressed to "each and all sons of the holy mother church" decreeing that "no master or bachelor of our Faculty (of Arts) should presume to determine or even to dispute any purely theological question. . ."[8] But there was needless to say no restriction on what the church may or may not deliberate or decide in matters logical or philosophical. The famous condemnation, under threat of excommunication, by Bishop Tempier of Paris, of 219 allegedly heretical propositions, ranged over such subjects as the (Averroist) doctrine of the double truth, the nature of primary matter and of first causes, the eternity of the elements and of the world, celestial motions, Zodiacal signs, free-will and the vacuum. Interestingly there were no condemned propositions that related to the nature of light or the science of optics. Possibly the study of this subject was not thought to present any temptations; or perhaps it also strengthened the power of the faith or the intellect to resist them? For some of the pious schoolmen of the thirteenth century, optics (*perspectiva*) represented even more!

The ultimate goal and steadfast purpose of the scholastic pioneers of optics was unequivocally affirmed: perfection of the soul. In principle the path to this goal was to be found in faith, in the divine revelation as enshrined in the sacred scriptures. But for frail mortals there was the need to translate revelation and faith into doctrine and practice. False doctrine was an ever-lurking danger: in the creed of the Scholastic there was no more worthwhile task than to establish the true doctrine, no duty more sacred than obedience to it. And to this end no source of guidance, inspiration or enlightenment need be excluded, *a priori*. To the contrary *all* available resources should be absorbed in this cardinal and boundless task. Historical circumstances seem to have

conspired to unite a fervent faith with devoted scholarship and ardent ratioci-
nation. Spiritual fathers and brothers, prelates, teachers and sages were at
once theologians, philosophers, and embryonic scientists: they were stirred
by a newly aroused interest in the workings of nature, in the Works of the
Creator as well as his word; and nourished by the newly acquired legacy of
Greece, Rome, Byzantium, and Islam.

Far from challenging the faith or true doctrine, the new learning with its
logic, reasoning and mathematics should result in their clarification and
strengthening. True piety could not reside in ignorance. Had not Maimonides
urged that "one who strives for human perfection must train himself first in
logic, then in mathematical disciplines . . .". Roger Bacon echoed this admonition, adding experimental science, metaphysics, and theology to the stern
curriculum.

The path would be long and arduous, not simply laid out before one. Divine truth, was not revealed at first glance. The true principles of God's design lay concealed beneath superficial appearances, to be revealed only by
patient probing, devoted scholarship, and deep understanding: was it not likewise with the pious understanding of his Word? The conjoining of the secular
with the sacred, in the service of a single overall truth, was an even more
challenging task. For many, the vast domain of nature, not to mention all
that had been written about it and was now accessible, was too much to grasp:
the new task of integration must, perforce, be exemplary rather than exhaus-
tive. And here, what better example could one choose than light itself —
transmitted through divine works and illuminating the truth through the divine word? The ardent Roger Bacon urges the matter thus:

> Divine truth, unrestrictedly considered must be understood and expounded;
> . . . this Science of Optics (Perspective) is necessary in both ways. For in God's
> Scriptures nothing is so much enlarged upon as those things that pertain to the
> eye and vision, as is evident to one who reads the scriptures through.[9]

How endless indeed are the opportunities for the pious investigator to em-
brace faith and reason in the study of light and illumination, of vision and
sight, light and darkness. Not drawing too clearly the boundary between allegory and analogy, between myth and metaphor, or between mystery and
mysticism, the medieval scholar could perceive divine illumination entering
the soul, analyzed by crystalline logic, reflected in the divine image, refracted
in human imperfection, and finally received by an instrument — the intellect,
for which, like the instrument of vision itself, we are indebted to a benevolent
Creator.

There were also prosaic tasks: How to formulate and articulate the new science, to ensure that it is fit and worthy of a place in the higher scheme of things. This was a task to which Bishop Grosseteste especially addressed his efforts; and he lay down a path for the guidance of others. The science of light must comprise two parts; the physics which was concerned with the actuality, the facts (*"quid est"*), and optics, perspective — which includes some mathematics — concerned with the explanation of the facts, of what is, (*"propter quid"*). The approach was metaphysical, for which Augustine provided the spiritual inspiration. The whole — the spiritual, the theological, the metaphysical and the mathematical — was provided with the underlying support of experience; not so much perhaps what Grosseteste himself had observed, but what he had learnt of the experience of others. On the whole, Grosseteste's optics offers precept rather than records practice. But there was no question as to the nobility or worthiness of its study: he placed optics highest, above all other sciences. He might well have chosen his text from the psalmist: "In Thy light we shall see the light."

There were perils as well as problems in this new intellectual domain. Could the pious accept with impunity the findings and teachings of pagans and infidels, who were uninspired and uninformed by true spiritual purpose or context? Would that not be to risk defiling the true faith, and to invite the accusation of heresy or blasphemy? Should not the guardians of the faith eliminate this risk by forbidding all contact with these works of the faithless? And so, early in the century (1210) all Aristotle's works and commentaries thereon were, at Paris, proscribed. A decade or two later (1231) the Lateran council proposed a more practical compromise to expurgate Aristotles' books of "anything virulent or otherwise vicious, by which the purity of the Faith might be derogated from. . .". In explication thereof a somewhat surprising analogy is drawn for the pious scholars; that "a comely woman who is found in the number of captives is not permitted to be brought into the house unless shorn of superfluous hair and of sharp nails."[10]

So, cleansed and shorn and with claws clipped!, the infidels and pagans Euclid, Aristotle, Ptolemy, Avicenna, Alhazan, and Averroës were admitted by the back door into the cathedral of faith. By mid-century, propagation of the wisdom of Aristotle — not without criticism and amendment — had become for Albertus Magnus a mission; and fifty years later, after the awesome efforts of Thomas Aquinas, Aristotle had become not only safe, but almost sacred.

Authority and justification for this assimilation of the profane to the sacred were not too hard to find. Had not the venerated Augustine taught that ". . . the gold and silver of the philosophers did not originate with them, but were dug out of certain mines, as it were, of Divine Providence."[11] There was no wisdom that was not contained in the wisdom of God: if pagans or infidels had been in possession of some useful part of it, could they not be deemed unworthy, unlawful possessors thereof? And in any case it was essential, as Bacon, urged, for *all* science to be pressed into service on behalf of the Church of God, and against the enemies of the Faith. Fortified by this assured self-righteousness, Bacon, like other scholastics from Grosseteste on, borrowed openly, usually with acknowledgement, but without reservation, from the Graeco-Islamic writings, and optics particularly from Euclid, Aristotle, Ptolemy, Avicenna (Abū ibn Siña) and Alhazan (Ibn al-Haitham).[12] Limits to this borrowing were imposed not by any scruple, but by availability. And so in substance medieval optics became largely a re-expression, with some reformulations, amendments and extensions of Islamic optics and of the Greek-Alexandrian science on which it was in turn based.

Two optical subjects commanded the special attention of the schoolman: one, vision and eye; the other, the rainbow. For both there was a ready spiritual context; and for both a substantial legacy of Graeco-Islamic writings. Much of this — the geometry of vision and the principles of reflection and refraction of Euclid and Ptolemy, the anatomy of the eye of Galen, and the basic scheme of transmission from luminous source to object and to the eye — was incorporated in the work of Alhazan (FIG. 1). The intrinsic nature of light remained an essentially speculative topic; and although Alhazan's own work had, in a *practical* sense, resolved the issue between intromission and extramission theories of vision, the metaphysical and theological speculations of the scholastics kept this ancient controversy verbally alive. Association of vision, light and illumination with the good, the noble and divine seemed inevitable. John Pecham could credit binocular vision to the "benevolence of the Creator (who) has provided there should be two eyes, so that if injury befalls one, the other remains."[13] A century earlier Alexandria Neckham had written regarding this subject.

> . . . how the two rays sent out to the eyes intersect each other transversely in the form of a cross; because just as without faith in the cross the inner man does not see well, so neither is external vision achieved without the form of a cross[14]

Roger Bacon takes his cue from the words of the psalmist: "Guard us O Lord as the pupil of Thine eye", and asks how it is possible " . . . to know God's meaning in his prayer unless one first considers how the guarding of the eye is effected, so that God deems it right to guard us according to this similitude. For when something is stated as an example of a similitude, that

FIGURE 1. Secular and sacred optical legend. Frontispiece to the first Latin edition of Alhazan's Optics: *Opticae Thesaurus Alhazini Arabis,* published by Frederic Risner, Basel, 1572. The same illustrates the edition of Witelo's Optics which is bound in the same volume.

At the bottom are depicted the optical themes of reflection and refraction. On the right a rainbow accompanies the end of the Deluge and the animals emerging from Noah's ark. On the left is shown the Roman fleet being set on fire in the concentration of sun's rays produced by an array of burning mirrors. Legend has it that Archimedes so defended Syracuse from Marcellus in 212 B.C.

which is exemplified cannot be understood unless the meaning of the example is also understood."[15]

It was then a duty to understand the structure and functioning of the eye; and for this it was necessary to study optics and apply mathematical reasoning, and through optical experiments to exploit the evidence of the senses to explore hidden causes. Only thus would the full import of the message be revealed; and in the meanwhile we may have learnt some practically useful mathematics and optics. This for Bacon was the full lesson: Science was necessary for the complete appreciation of faith: it may also be useful in its defense. The task would not only be rewarding, it was also demanding: "Science is acquired at great expense and by frequent vigils, by great expenditures of time, by sedulous discipline of labor, by relevant application of the mind."[16]

The rainbow was equally striking as a manifestation of God's handiwork and as a challenge to the human intellect. From the time of Aristotle to the end of the thirteenth century, it represented perhaps the greatest challenge to optical science: the perfection of its circular arcs beckoned the geometer with his irrefutable logic; the spectacle of color stimulated conjecture and experiment with crystal and water; the varied iridescence and illusion with sun, rain, clouds and observer invited an endless host of observations and interpretations. "The end of true philosophy" wrote Roger Bacon "is to arrive at a knowledge of the Creator from knowledge of the created world." Elucidation of the rainbow seemed clearly directed towards this goal. The spiritual purpose was declared explicitly: "I do set my bow in the cloud and it shall be a token of covenant between me and the earth" (Genesis 9:13). But the full import of this message might lie deeply hidden, only to be revealed with the aid of scientific understanding.

The major aspects of the rainbow's appearance and the significant times and circumstances of its occurrence had been succinctly outlined by Aristotle, 1500 years earlier.[17] He had also indicated the salient geometrical features, and had attempted an interpretation, of sorts, of these. "Aristotle," commented Grosseteste, "in his book on meteorology has not revealed the explanation as it concerns the student of perspective, but he has considered the facts of the rainbow which are of concern to the physicist. . ." From Aristotle to Alhazan, there had been occasional added observation, but little real progress in understanding. Ironically the far better understanding of the ordinary visual process achieved by Alhazan, had helped little, had even perhaps hindered, a deeper insight into the subtlety of the rainbow apparition. But now in the thirteenth century the attention of the pious schoolmen turned again to this marvel of God's handiwork, and assisted by the whole legacy of Aristotle and

Islam, and guided by their own faith in both experiment and reason, significant advance was made at the end of the century.

Although Grosseteste recognized, in principle, that more facts, as obtained by physicists and not necessarily concerning only the rainbow itself, might help the student of optics in his interpretation of the rainbow, and although he did himself add some dubious observations which ostensibly refuted Aristotle's theory, his own theory can hardly be regarded as more successful. Roger Bacon attacked the problem with the full fervor and confidence of a champion of the new experimental science, and especially of its most "noble" application in the field of optics. He reassembled all the many facts (and some fictions!) that had been garnered and recorded over the years; and added some exemplary material of his own. Following the cue of the rainbow colors (FIG 2), he sought out and scrutinized all occasions where such colors arise — from sunlight falling upon crystals and prism, on fountains and falling rain, on the spray from an oarsman's blade and from the mill-wheel, from dew-drops in the grass and in spider's webs, in half-closed eyes, "and the diligent observer may find many more examples."[18] But he was not convinced, as was Grosseteste, of the reality of these colors. Perhaps they were in some sense illusory, produced in the human eye? And did their similarity imply a common cause? Perhaps not, since the colors in the crystals seemed to be located therein, whereas, as he observed, the rainbow arc and colors appeared to move with the beholding eye. Bacon wrestled with these problems — inconclusively. Turning to the geometry of the bow, and accepting the basic scheme of Aristotle, with an astrolabe he made the first recorded measurement, 42°, of the angle of the cone of which the rainbow arc subtends a segment at the observers eye. He sensed, but could not resolve, the strange subjectivity of the rainbow image, and other illusions; and finally he confessed:

> Hence reasoning does not attest to these matters, but experiments on a large scale made with instruments and by various necessary means are required. Therefore no discussion can give an adequate explanation of these matters, for the whole subject is dependent on experiment. For this reason I do not think that I have grasped the whole truth; because I have yet made all the experiments that are necessary; and therefore in this work I am proceeding by the method of persuasion and of demonstration of what is required in the study of science, and not by the method of compiling what has been written on this subject. . .[19]

In sum, Bacon modestly declared his work to be only a "plea for the study of science"!

Bacon had also, following on some speculations of his mentor Grosseteste, spent much effort in the exploration and analysis of vision through the layers

FIGURE 2. Traditional optical themes. Frontispiece illustration to a 1551 edition of Witelo's Optics. Shown are the burning glass, reflection, refraction and the rainbow. The collinearity of the sun, the observer's eye, and the center of the bow are correctly displayed.

of refracting material, with both plane and spherical boundary surfaces. He had observed, and attempted to explain, the magnifying power of a segment of a glass sphere. He shared Grosseteste's conviction that

> this part of optics when well understood, shows us how to make things a very large distance off appear as if close and large, and near things very small; and how we make small things placed at a distance appear any size we want. . .[20]

It is possible that these observations had some bearing on the mysterious appearance of spectacles at the end of the thirteenth century. Bacon, to be sure, was not insensitive to the possibilities of applied science: "What nature can do, Art perfecting Nature can accomplish in greater measure."[21]

Albertus Magnus' contribution to optics was but a small part of his vast erudition, which ranged over all the sciences, philosophy, metaphysics, ethics, and politics and embraced the task of making the whole Aristotelian corpus accessible and intelligible to the Christian, Latin west. Of his spiritual purpose, there seems to be no question; but God, he asserted, acts through natural causes, and he did not therefore presume to seek the Divine Will directly, but rather by investigating the causes themselves, *i.e.* through natural sciences. Moreover, it was "the aim of natural science not simply to accept the statement of others, but to investigate causes that are at work in nature."[22]

But, once again, it seems to have been as much precept as practice. For Albertus, nature was revealed by *"experimenta"* which could denote experience generally, included in the reports of others, or by *"experimentation"* which suggests direct experience of his own. He had a tendency to use the two terms interchangeably. His views on vision reflected those of Avicenna and Averröes. In his discussion of the rainbow he went little beyond Grosseteste, but he stressed as did the ancients, the role of falling rain-*drops* as distinct from the cloud as a whole, and he claimed to have observed colors in the shadows cast by raindrops in sunlight. In the spirit of Aristotle he illustrated the notion of color being intermediate between light and darkness: White sunlight passing through transparent water, is rendered red by the addition of black ink. (Echoes of Aristotle's red wine from black grapes!) Albertus Magnus was little closer than his predecessors to perceiving how the effects of innumerable individual drops could combine to produce the overall rainbow spectacle.

Witelo's optics derived largely from the work of Alhazan, and although it lacked explicit acknowledgment thereto, there were borrowings from Grosseteste and Bacon. He examined further the phenomenon of refraction; and stressed its role, in addition to reflection, in the production of the rainbow arc and colors. Witelo had in his hands most of the optical pieces necessary to resolve the rainbow mystery; but his analysis was too crude — the subtlety was deeper than he suspected. It fell to a third member of the Dominican order, Theodoric of Freiberg, finally to achieve a really significant measure of success.

One cannot but wonder how a leading Dominican of his day, in the face of all his administrative-spiritual tasks and obligations, could find time and opportunity to write innumerable tracts on Logic, Natural Philosophy, Cosmology, Theology and Ethics, included in which were treatises on light and color and by far the most detailed study, up to that time, of the rainbow (and associated "radial impressions"). Theodoric's *De iride* embraced the full range of experiment and observation, theory and conjecture: it represents the most significant step towards the full rational explanation of the rainbow mystery since Aristotle. It has been enthusiastically hailed by modern writers as "perhaps the most perfect use of the experimental method during the century that

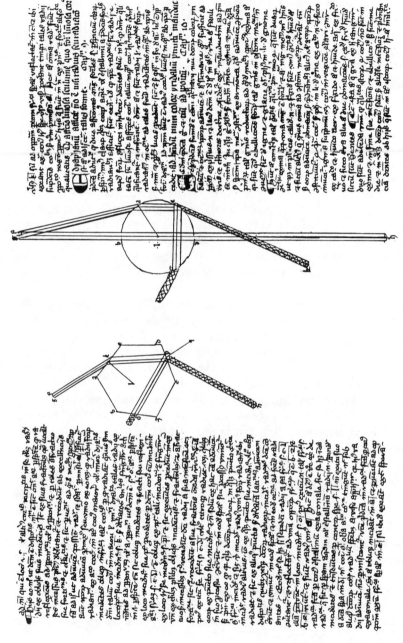

FIGURE 3. Theodoric's experiments with prism and sphere. Showing the similarity of refracted transmission and reflected transmission with refraction and internal reflection, in the two cases. In both sunlight is falling from above. The cross-hatching indicates that both transmitted and reflected deviated rays carry color. The latter correspond to the rainbow light.[7] (Reproduced from a manuscript of *De iride* at the University of Basel.)

followed Grosseteste's work . . . A model example of the thirteenth century theory of experimental science in practice, and a model of experimental procedure for all time."[23] Even if it falls somewhat short of this lofty hyperbole, Theodoric's singular achievement would still command close attention (FIG. 3).

The provenance of Theodoric's scientific methodology and optical knowledge seems fairly clear. Aristotle was by far his most important, if not-always-followed, guide: references to the Philosopher abound; those to Euclid and Ptolemy and Proclus are less frequent, to Plato himself rare; the Church Fathers Augustine and Boethius were constant sources of inspiration. At the time he was writing, a great bulk of Arabic works was available in Latin. Theodoric leaned heavily on the optical writings of Alhazan (referred to as *"auctor perspectiva"*) and to a lesser extent on those of Averröes (*"Commentator"*), Avicenna, and Alfaribus. Although there are no references to most thirteenth-century writers, there is little doubt about the influence of Grosseteste, Bacon, and Witelo on Theodoric's work. In all, Theodoric's optics was decidedly eclectic. "How hard it is" he declared in the preface to *De iride*, "on this well-trodden path, to find new ways, detours and shortcuts." There is no question about the respectability of Theodoric's activity: the treatise on the rainbow was written at the expert request of, and is dedicated to, the Master General of the Dominican Order of Preachers, Aymeric de Plaisance.

This was one of the last of Theodoric's works; for which the earlier writings on light (*De luce*) and color (*De coloribus*), together with earlier extensive philosophical and metaphysical works (especially *De accidentibus*) provided a general theoretical basis. The correct note of reverential spirituality was struck at the outset in the treatise on Light: "In what way is the light divided and the east wind scattered over the earth?" (Job 38: 24). Theodoric recalled that this difficult question was one of many others, equally grave, which God put to Job. The whole sequence on light, color, the rainbow and other like phenomonena was presented by Theodoric as a logically ordered sequence, moving from general principles to specific problems, with the final deductive interpretation of phenomena "confirmed" by summary comparison with experiment and observation. There is not much doubt about the order in which the parts of the work were written; but it is quite another matter to surmise in what order the knowledge contained therein was acquired and the opinions and conclusions formed. Experiment and observation usually follow theoretical argument, which assumes thereby the guise of prediction; but as often as not, this sequence is hardly possible. Theodoric undoubtedly made many experiments and observations, including some original; but exactly what and how they were made and what was directly inferred from them is not always clear from the very sparse description. Though basically Aristotelian and qualitative, Theodoric was not insensitive to geometrical-mathematical aspects of his subject; but rarely in his experiments did he record anything quantitative of his apparatus or observations (not uncharacteristic of medieval science), so that even a detailed examination of his work leaves many unanswered questions.

Theodoric's analysis of the nature of light followed typically Aristotelian–Scholastic lines — the usual classification of formal, material and efficient causes. His "theory" of color was somewhat more novel. Basically following Alhazan, and in contrast to Bacon, he treated color as real, as real as light itself. Of the three different modes of producing colors, the most interesting was that of "radiant colors" such as observed in the passage of sunlight (or viewing the sun) through a colorless crystal (glass) prism. To Theodoric these prismatic, colored effects could be considered as "images" of the luminous body; and just as reflected images in mirrors were not exact replicas of the objects, so also the colored images produced by refraction in prisms were imperfect replicas. This repeated comparison of refraction and reflection may also give some hint of how Theodoric surmised, or discovered, that light passing from a dense medium (*e.g.* glass) to a rare medium (*e.g.* air) is not all transmitted, but is, in part, reflected. This was his most original new physical principle, which he later introduced into the interpretation of the rainbow. The process by which colored images could be produced was analyzed in terms of his metaphysical principle of "contraries," and a theory was elaborated which yielded *four* radiant colors; four being the product of two quasi-formal contraries and two quasi-material ones. And "four" radiant colors — red, yellow, green, blue — were what Theodoric claimed one observed as a *fact*. This "fact" of four colors, in contradiction with Aristotle's assertion of *three* rainbow colors, provided Theodoric with an opportunity to rebuke the Philosopher with his own advice that "one should never renounce the manifest information of the senses!" Theodoric could find additional support for his fourfold, two-times-two, scheme of colors, in a comparison with Empedoclean-Aristotelian four-element scheme; and he drew the parallels: red–fire, yellow–air, green–water, and blue–earth!

Presumably it was in connection with this analysis of color that he made many of the recorded observations of glass prisms and globes. He recorded the observations with hexagonal prism: that passage of sunlight through the refracting material did not, *per se*, produce color: deviation is also necessary. He observed here the partially internally reflected light, as well as the transmitted rays, and that these two yield colors. He examined the colors either by allowing the shaft of sunlight to fall on a screen, or by viewing the sun through the prism; and recorded changes with changing distances between

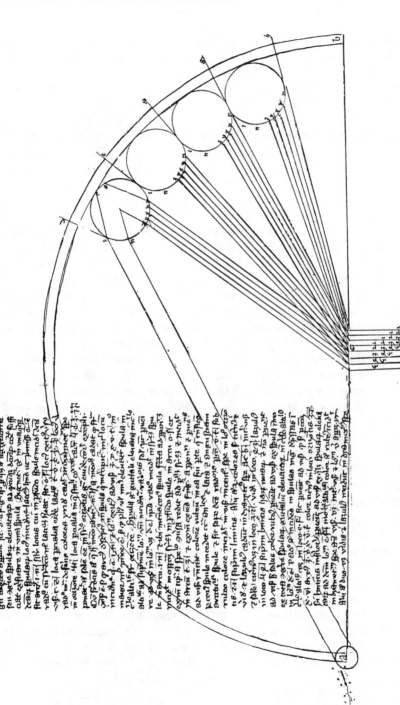

FIGURE 4. Theodoric's primary (lower) rainbow. Here the sun is shown on the horizon at the far left (a). The observer is at the center, and on the right are, grossly magnified, the spherical raindrops. The different emergent rays that may reach the observer are marked t,v,x,y,z, representing the boundaries between the four different colors.[7] (Reproduced from a manuscript of *De iride* at the University of Basel.)

the prism and eye, or screen. The observations, as recorded, do not suggest very searching experimentation, nor do all of them seem wholly plausible. Nonetheless their interpretations all provide what Theodoric believed to be further demonstration of the veracity of his color theory (FIG. 4).

Turning to the rainbow and like atmospheric phenomena, Theodoric argued that there were five possibly relevant, basic optical processes: (i) simple reflection on the external surface of a sphere (raindrop); (ii) transmission, with refraction on entry and exit; (iii) refraction, internal reflection, and refraction again on exit; (iv) the same two refractions but now with *two* internal reflections; and (v) refraction in an extended cloud/mist of varying density. He showed that (iii) and (iv) were the keys to the rainbow image. Whether Theodoric was led to examine the reflected light from a sphere to test the analogy with the similar process with the hexagonal prism, or whether the sphere experiments were inspired by the many earlier accounts of them, it seems likely, in either case, that his attention was early drawn to the striking and brilliant reflected colored images, visible when one looks toward the edge of the globe, in the direction of incoming sunlight. He recorded seeing in the globe the standard *four* radiant colors! With the hexagon, the reflected colored light is at a definite angle to the incoming light (since the prism has planar faces); that there should be a similar definite angle for the sphere is perhaps surprising, but Theodoric recorded neither this specific fact—nor surprise. In any event, independently of any *interpretation* of the alleged four colors, the *observation* of the bright reflected light in the glass globe with the same sequence of colors as in the rainbow would surely have been a powerful a clue as any theoretical proposition. It was on this correspondence between sphere and rainbow, whether theoretical or observational, that the rainbow interpretation rests (FIG. 5).

Accurate measurement and cogent geometrical argument were not Theodoric's forte. Just how the two phenomena are related is essentially a geometrical question of modest subtlety—provided each phenomenon is separately described correctly and fully. But Theodoric's picture of the heavens was a medieval hemispherical canopy with the sun and the clouds at equal distances from the centrally located observer! An essential feature of parallel sun-rays is missing! But Theodoric seemed unaware of the significance of the *direction* of the light in both globe and rainbow phenomena: repeatedly he confused angle and position and compounded error and ambiguity. Surprisingly, he stated, and repeated, the critical angle of the bow at the observer's eye to be 22° (not 42° !), and asserted that this "can be measured by an astrolabe." "Can be," notice, not "has been"! It was a familiar style of the day and Theodoric's normal mode of expression: "I say that. . ." was the common form

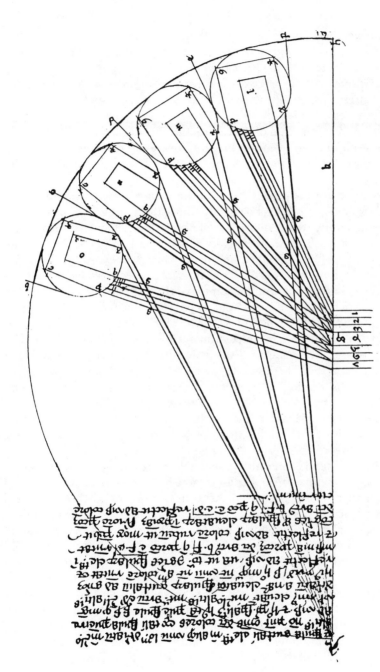

FIGURE 5. Theodoric's secondary (upper) rainbow. The same configuration as FIGURE 4. Two internal reflections inside the drop now result in a cross-over of the incident and emergent rays; and this is associated with the reversed order of the rainbow colors.[7] (Reproduced from a manuscript of *De iride* at the University of Basel.)

of recording an observation, not "I saw that. . ."! He did give the correct angular separation between primary and secondary rainbow: and this is perhaps easier to measure than either bow itself.

Theodoric's theory is a sort of caricature of the rainbow — blending reality with the impressions, the views and predictions of its draughtsman. But in its major features it is undoubtedly recognizable: and these features appear in the "correct" interpretation of the future. Of Theodoric's theory of "radiant colors" we seem to hear no more. Exactly how, and how much theory and observation, speculation and experiment guided Theodoric, in making his "solution" of the rainbow puzzle must clearly remain a matter of conjecture. We may guess that more often than not observation came first, and that its ample folds of ambiguity made it possible to drape theory over facts, *a posteriori*.

Unknown, surely, to Theodoric, at almost exactly the same time, thousands of miles away, in distant Persia, the same basic conclusions were revealed by Qutb al-Din al Shirazi, and his disciple Kamal al-Din al-Farisi.[24] Theirs was no metaphysics of color, but the same reflections and refractions in the individual raindrops, and the same basis for explanation: observations of the passage of sunlight through, and its reflection by, the waterfilled globe (FIG. 6). Kamal al-Din's experiments seem more detailed and cogent; he attempted to interpret them by calculating the passage of different individual rays through the globe — using the Ptolemaic-Alhazan knowledge of air-water refraction. By contrast Theodoric made no use of this quantitative knowledge of refraction. Kamal al-Din seemed to recognize that the unique angle of reflection from the globe was a real problem, and in his ray analysis he seemed close to recognizing that this results from critical maximum/minimum angles of deviation. Theodoric's interpretation of the narrow band of the rainbow fell back essentially on Aristotle's arbitrary assumption.

What Theodoric and Kamal al-Din shared was a common legacy of Islamic-Greek science, and Alhazan's optics especially. Each was moved to extend and correct his master's work: Theodoric, Aristotle's and Kamal al-Din, Alhazan's. The thirteenth century "solution" of the rainbow mystery was the end rather than the beginning of a chapter in the history of optics. It was an exploitation of old ideas, rather than the discovery of new. Unlike Ptolmaic–Alhazan optics, which was transmitted by Witelo, Pecham *et al.*, and was widely read for two or three centuries at least, Theodoric's work made little impression. Johannes Kepler's celebrated *Foundations of Optics*, of 1604 is subtitled *Supplement to the Optics of Witelo*. From the fourteenth century to the end of the sixteenth century there was little advance in optics, and what few attempts there were to explain the rainbow seemed to ignore experiment, including the experimental discoveries of the medieval scholars. When René Descartes, in 1637,

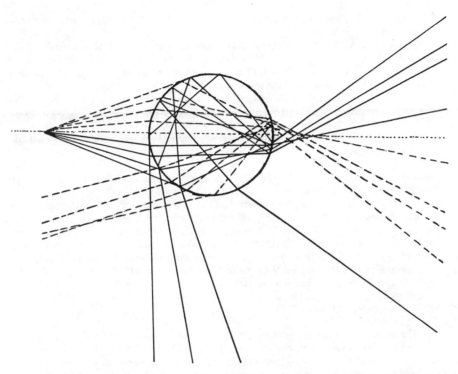

FIGURE 6. Ray-tracing in a transparent sphere by Kamal al-din al-Farisi. The (theoretically calculated) rays are shown as diverging from a point-like luminous source at the left. Once-reflected, and partially transmitted, rays are shown by dashed lines; twice-reflected rays by solid lines.

presented his definitive explication of the geometry of the rainbow arc, he made no mention, needless to say, of Theodoric of Freiberg. Kamal al-Din's work was not "discovered" in the west until late in the nineteenth century.

NOTES AND REFERENCES

1.   Alexander Neckham (also Neckam) 1157–1217. "Optics," in the encyclopedic *De naturis rerum* (On the Nature of Things), *ca.* 1200.
2.   Robert Grosseteste, *ca.* 1168–1253. (a) Main writing on physics and optics in *De motu corporali et luce, De speculis!* 1215–1230; Commentary on Aristotle's Physics; *De iride; De colore.* 1230–1235. (b) For a full appreciation of Grosseteste's work: A.C. Crombie, *Robert Grosseteste and the Experimental Method* (Oxford: Oxford University Press, 1953). (c) For a convenient sample of extracts of writings on optics, together with brief biographical notes, by Alhazan, R. Grosseteste, Roger Bacon, John Pecham, Witelo, Theodoric of Freiberg *et al.*,: David C. Lindberg in *A Source Book in Medieval Science.* E. Grant, ed. (Cambridge, MA: Harvard, 1974). (d) For

authoritative biographies see *Dictionary of Scientific Biography.* C.C. Gillispie, ed. (New York: Charles Scribner's Sons, 1980).

3. Roger Bacon, *ca.* 1219–1292. Main optical writings in *Opus majus, ca.* 1266. English translation: R.B. Burke, 1962 edit. (New York: Russell and Russell). Also David C. Lindberg, *Roger Bacon's Philosophy of Nature* (Oxford: Oxford University Press, 1983). Introduction contains a short review of the history of light metaphors.

4. Albertus Magnus, *ca.* 1193–1280. (a) Commentaries on Aristotle's Physics, *De causis et prop. elementorum, meteora,* etc. 1245–1260. (b) *Albertus Magnus and the Sciences,* J.A. Weisheipl, ed. (Toronto: Pontifical Institute of Medieval Studies, 1980).

5. Witelo (also Vitellio), *ca.* 1230–1275 Optical writings in *Perspectiva, ca.* 1270 Printed Editions 1551, 1572, etc.

6. John Pecham (or Peckham), *ca.* 1235–1292. *Perspectivia communis, ca.* 1275. Printed Edition, Georg Hartman, ed. 1542. Also *John Pecham and the Science of Optics,* (Engl. translation) David C. Lindberg, Trans. (Madison: University of Wisconsin Press, 1970).

7. Theodoric of Freiberg (also Meister Dietrich, Thierry de Friburg) *ca.* 1255–1315. Natural Philosophy: *De natura contrariorus; De elementis corpora naturalis,* etc. Optics: *De luce et eius origine; De coloribus; De iride,* 1306–1310. *De iride,* edited by J. Würschmidt, *Beiträge zur Geschichte der Philosophie des Mittelalters* 12 (1914): 33–204; also W.A. Wallace: *The Scientific Methodology of Theodoric of Friedberg* (Fribourg, Switzerland: University of Fribourg, 1959).

8. Translated by Lynn Thorndyke: *University of Records and Life in the Middle Ages* (New York, 1944). Quoted by E. Grant in reference 2(c). p. 44.

9. *Opus majus,*[3] vol. II., p. 576

10. Thorndyke,[8] p. 43.

11. *Opus majus,*[3] p. 634.

12. The major works relating directly to optics are: Aristotle (384–322 B.C.) *Physica, Meterologica,* etc. Translated from Greek and Arabic, 12th century. Euclid (*ca.* 330–260 B.C.), *Elements, Optica.* Latin translation from Greek and Arabic, 12th century. Ptolemy (mid-2nd century A.D.), *Optica (De aspectibus),* Latin translation from Arabic, *ca.* 1150. Avicenna (980–1037), *Kitab al-Shifa (De iride),* Latin translation from Arabic, 12th century. Alhazan (*ca.* 965–1039). *Kitab al Manazir (Perspectiva/de aspectibus)* Latin translation from Arabic, end of 12th century. Latin Edition, edited by Frederic Risner, Basel, 1572; *Cf.* A.C.Crombie, *Medieval and Early Modern Science* (Cambridge, MA: Harvard University Press, 1963).

13. *Perspectiva Communis*[6] Prop. 32. Quoted by Lindberg,[2c] p. 399.

14. *De naturis rerum*[1] Quoted by Lindberg,[2c] p. 382.

15. *Opus majus,*[3] p. 580.

16. Neckham,[1] Quoted by Lynn Thorndyke, *History of Magic and Experimental Science,* vol. 2 (New York: Macmillan, 1923), p. 196.

17. Aristotle's *Meterologica* was translated directly from the Greek in the mid-twelfth century. See reference 12.

18. *Opus majus,*[3] p. 590.

19. *Ibid.,* p. 615.

20. *De iride,* quoted by Crombie,[2b] p. 123.

21. *Opus majus,*[3] p. 580.

22. Albertus Magnus *De mineralibus.* Quoted by J.A. Weisheipl.[4b]

23. Crombie,[2b] p. 233.

24. Kamal al-Din al-Farisi, *Tankih al-Manazir,* ca. 1304/1310. German translation; Revision of Alhazan's Optics. *Physikalisch-medicinische Societät zu Erlangen. Sitzungsberichte* 42 (1910): 15. Also, Wallace.[7]